Roman Brick Stamps in the Kelsey Museum

The University of Michigan
KELSEY MUSEUM OF ARCHAEOLOGY
Studies 6

ROMAN BRICK STAMPS IN THE KELSEY MUSEUM

By John P. Bodel

Ann Arbor   The University of Michigan Press

Library of Congress Cataloging in Publication Data

Bodel, John P., 1957–
    Roman brick stamps in the Kelsey Museum.

    (Studies / Kelsey Museum of Archaeology ; 6)
    Bibliography: p.
    1. Brick stamps—Rome—Catalogs. 2. Brick stamps—
Michigan—Ann Arbor—Catalogs. 3. Inscriptions, Latin
—Catalogs. 4. Kelsey Museum of Archaeology—Catalogs.
I. Kelsey Museum of Archaeology. II. Title.
III. Series: Studies (Kelsey Museum of Archaeology) ; 6.
CN529.T5B62   1983   929.9   82-24841
ISBN 0-472-08039-3

UXORI CARISSIMAE

CONTENTS

PREFACE

This study will be read mainly by specialists, but I have tried to present the material in such a way as to be intelligible, and perhaps even useful, to the general classicist. It is not intended as a general introduction to the study of brick stamps, but I hope that English readers with an interest in the social and economic history of the Roman Empire will find my comments an aid in approaching what is perhaps the most intractable of all inscriptional evidence. I hope also that brick stamp experts drawn by the prospect of encountering new material will not be put off by the general remarks meant for the non-specialist. The primary purpose of this edition, after all, is to make available to scholars a sizeable body of material that would otherwise have remained inaccessible, and indeed, whose very existence has until now gone unrecognized.

In the course of my research I have turned to many for advice and practical help, and it is a pleasure to record here my gratitude to those whose assistance has been especially instrumental in shaping this monograph into its present form. My warmest thanks go to John G. Pedley, Director of the Kelsey Museum, not only for permitting me to publish the Michigan collection of examples, but also for ensuring that my work appear in the series of Kelsey Museum of Archaeology Studies and for providing steady encouragement throughout the many phases of my research. I am deeply obliged to my advisers at the Department of Classical Studies, Professors J. H. D'Arms and B. W. Frier, who read the whole work in draft and enabled me to make a number of improvements in content and presentation; the Introduction in particular owes much to their thoughtful criticism. The extent of my debt to Professor Margareta Steinby will be evident from the frequency with which her name appears in the following pages, but I should like to express particular gratitude for the careful attention she devoted to an early version of my manuscript: thanks to her generosity I have changed the text in literally dozens of places. If in the comments I have acknowledged my obligation wherever her influence has been most direct, I must mention here the generally salutary effect that her correspondence has had in guiding my decisions in editorial matters. I have profited also from conversation with Professor Silvio Panciera, who at a late stage discussed with me certain persistent problems and pointed the way to some possible solutions. Whatever errors or misconceptions remain are entirely my own.

Credit for the excellent photographs is due to Mr. F. Anderegg, whose skillful execution can be appreciated the more readily if one compares his work with the few photographs that for one reason or another have had to come from other sources. I am grateful to David Slee for drawing the artwork and for helping arrange the plates, and to Jeffrey Henderson, who read the proofs, for saving me from many blemishes. Finally, I wish to express my deep sense of appreciation to Bruce Frier, who suggested the project to me initially and who helped its progress at every stage; his consistent and friendly support has been the more valuable to me for being unobtrusive. And to my wife, Helen, I owe an enormous debt of gratitude: her remarkable patience and understanding have contributed more to the successful completion of this book than she can ever know; it is as a small token of my appreciation that I dedicate the following pages to her.

Ann Arbor   J. P. B.
July 1982

I. THE STUDY OF BRICK STAMPS

Stamped bricks were used in ancient Egypt already under the Theban kings and in Greece from at least as early as the fourth century B.C., but Egyptian and Greek brick stamps, relatively few in number, have as yet received little scholarly attention; instead, the study of brick stamps has centered on the Latin inscriptions that first appear on Roman bricks during the last century of the Republic. At first the inscriptions were simple—a single name, usually, stamped into the wet clay with a die made of hardwood or, occasionally, metal. But when burnt brick became the preferred building material at the capital in the early Empire and wealthy landowners throughout the lower Tiber valley began more and more frequently to exploit the claylands on their estates for brick production,[1] the number of distinguished names appearing on brick stamps multiplied, and the stamps themselves grew more complex. By the time of Hadrian the texts often recorded not only the names of the *figlinae* where the brick originated and the owner of the land, but also the name of a second person, the *officinator*, and, through the paired names of consuls, the year in which the brick was made.

This wealth of information distinguishes Roman brick stamps from the stamps found on amphorae, lamps, and other smaller utensils, and also, for the most part, from the brick stamps used elsewhere in Italy and the provinces.[2] In the past this information has attracted the interest primarily of prosopographers and building historians, for the names encountered on stamps are frequently those of men and women prominent in the political history of the period; and since most of the stamps are at least approximately datable, they can often be used to date the buildings in which they occur. More recently, social and economic historians have begun to take advantage of a characteristic of Roman brick stamps that sets them apart from other types of inscriptional evidence: for many purposes, the body of material is virtually complete. So many hundreds of thousands of ancient Roman bricks have survived that today we know almost all the different stamps that there ever were. As an epigraphic source, the Roman brick stamps are thus peculiarly suited both for statistical analysis and for case studies; they enable us on the one hand to examine comprehensively the involvement of various social groups in Rome's largest industry, and on the other to follow in detail the development of individual careers.

The importance of brick stamps as historical documents was evident to the first editor of the *instrumentum domesticum* of Rome, Gaetano Marini. In his *Iscrizioni antiche doliari*, Marini furnished the texts of nearly 1500 stamps with individual comments that are characterized above all by astute prosopographical observations, many of which have since found general acceptance; his accuracy in reading the texts, moreover, though questioned at times by later editors, is continually being confirmed by new discoveries.[3] Completed in 1799, Marini's work remained in manuscript for nearly a century; when it was

---

[1] We need no longer adopt the traditional view that brickmaking was considered respectable activity for Roman senators only because it constituted a form of agriculture: see J. H. D'Arms, *Commerce and Social Standing in Ancient Rome* (Cambridge, Mass. 1981), p. 156 *et passim*; hereafter cited as D'Arms, *CSS*.

[2] The same stamps found on bricks are, however, sometimes found also on certain large clay products—*dolia* and *mortaria* in particular, but also sarcophagi and even a few terracotta reliefs. For consular-dated brick stamps used in Italy outside Rome, see H. Bloch, *AJA* 63 (1959), p. 235 n. 40. Features peculiar to the stamps made in the capital and its vicinity are terms such as *praedia* and *opus doliare* (the latter occurs only in brick stamps); mention in the text of two persons; the use of figured symbols, *signa*, other than for purely decorative purposes; and especially, the "orbicular" form, or one in which a small circle, the orbiculus, intrudes into the circumference of a larger circle. (◉).

[3] Cf., for example, the Catalogue numbers **4** and **98**. Entries in the Catalogue are hereafter indicated by numbers in boldface type without further specification.

finally published in 1884 by G. B. De Rossi, it came equipped with an index by Giuseppe Gatti and supplemental notes by a pupil of Theodor Mommsen's, Heinrich Dressel.

Dressel was at the time compiling a new edition of Roman brick stamps for the *Corpus Inscriptionum Latinarum* in which the material was to be arranged by the following categories: public stamps; private stamps from named *figlinae*; stamps of emperors; unattributed private stamps, including those of the *gens Domitia*; stamps from the age of Diocletian or later; broken or poorly reported stamps; and "suburban" stamps from the outlying areas of the Roman Campagna.[4] Based primarily on place of discovery rather than place of manufacture, Dressel's classification of stamps as "urban" or "suburban" was in many cases misleading; and just as the hundreds of new stamps that have come to light since the publication of *CIL* XV,1 in 1891 need to be incorporated into his system, so also reevaluations of material already known to Dressel have recommended substantial changes in his arrangement of the texts. Thus, *CIL* XV,1 is out of date and the need for a new edition is patent. But until a comprehensive review of all the extant examples can be successfully attempted,[5] Dressel's volume, now nearly a hundred years old, will remain the standard edition of Roman brick stamps, and his organization of the material will continue to serve as the inevitable point of departure for any new interpretation.

In *CIL* XV,1 Dressel assigned to any stamp not marked with the consular year an approximate date which he usually based on the following criteria: first, mention in the text of persons known and datable from other sources, including other brick stamps; and second, the shape of the stamp and the character of the lettering. The second criterion he derived directly from the first. By comparing stamps that could be dated from the persons appearing in their texts, Dressel was able to trace the developments in form that brick stamps underwent during the first and second centuries A.D.[6] Although the chronology of types that he worked out on these principles has frequently since been corrected and refined, the general validity of his procedure is today universally recognized. Furthermore, in reporting all the known examples of each stamp and in reproducing the texts with a degree of accuracy that remains unparalleled, Dressel established a solid foundation for further research. But it was not until the 1930's that one scholar was able to treat brick stamps as more than objects of limited interest only to specialists, and was able conclusively to demonstrate their importance not only for the building history of the capital but also for the social and political history of the early Empire.

In his pioneering study, *I bolli laterizi e la storia edilizia romana*, Herbert Bloch established once and for all that brick stamps provide a reliable means of dating the buildings in which they are discovered, often to within five years or so of the actual date of construction.[7] When bricks found with some regularity in securely dated structures were then found also in otherwise undated buildings, they were shown to date not only the buildings themselves but also any other stamps occuring frequently in the same architectural context; these newly dated stamps could then be used to date other buildings and stamps, and so on. The precision that Bloch achieved through his new system of dating, and the profound impact that the successful practical demonstration of the method had on the history of public building at Rome and Ostia are generally considered to be his most important contributions to the study of brick stamps.[8] But his accomplishments in the areas of prosopography and social and economic history are hardly less

---

[4] In a second section Dressel assembled the stamps found on *dolia; mortaria*, which he incorrectly called *pelves; arcae; tubi; antefixa;* and *laterculi anaglypti.*

[5] A thorough revision of the material would still today be premature if not impossible since many collections remain unpublished, while others must be reexamined in the light of recent developments in the standard of precision expected of editors of brick stamps: see the remarks of M. Steinby in *BullCom* 86 (1978-1979), p. 55, and below, p. 4.

[6] Dressel first presented his methodology in an important study, *Untersuchungen über die Chronologie der Ziegelstempel der Gens Domitia* (Berlin 1886), which he later summarized in *CIL* XV,1, pp. 265–275; see also the introduction to *CIL* XV,1, pp. 9–10 where the criteria for dating stamps in the edition are concisely stated.

[7] Bloch's work first appeared in the *Bullettino Comunale*, vols. 64–66, in the years 1938–1939 and was later reprinted, with indices, as a single volume (Rome 1947 and 1968). Page references are to the book edition, which is hereafter cited as Bloch, *BL.*

[8] For the methodology, see *BL*, pp. 341–343. Practical demonstrations: *BL, passim*; "I bolli laterizi e la storia edilizia di Ostia", appendix in *Scavi di Ostia* I (1953), pp. 215–227; "Sette Bassi Revisited", *HSCP* 63 (1958), pp. 401–414; and perhaps the most striking instance, "The Serapeum of Ostia and the Brick-Stamps of 123 A.D.", *AJA* 63 (1959), pp. 225–240, wherein it is shown that buildings abounding in stamps from the years 123–125 were built immediately thereafter, precisely in the years 126–128.

consequential;[9] and with his *Supplement* and *Indices* to *CIL* XV,1, Bloch made the body of material readily accessible to those not already expert in the field.[10] The *Indices* are an invaluable aid in identifying fragmentary examples, and they greatly facilitate any systematic investigation of the evidence; the *Supplement*, quite simply, is the best English introduction to the study of brick stamps.

Six years after Bloch's last articles on brick stamps appeared in 1959, a team of Finnish scholars began cataloguing and analyzing the brick stamps of Ostia. Their research led in the 1970's to the publication of three monographs, by Tapio Helen, Margareta Steinby, and Päivi Setälä, on the organization, chronology, and private *domini* of the Roman brick industry.[11] Of these, the most valuable is that of Steinby. Like Bloch, Steinby was chiefly concerned with dating, and her study is an authoritative chronology of the urban *figlinae* active during the first two and a half centuries of Empire.[12] But Steinby's argument, like Bloch's, embraces broader historical issues as well; her remarks are full of interesting observations, particularly on the use of *signa* as distinguishing emblems and on the connections, familial and professional, between persons involved in the production of bricks.[13]

Helen's work is more controversial. According to Helen, the term *figlinae* refers not to a production unit (brickworks), as was previously supposed, but rather to a territorial district (claylands); *domini*, then, were not manufacturers, but landlords; *officinatores* were not their subordinates, but were themselves independent entrepreneurs. In transferring the initiative in brick production from the *dominus* to the *officinator*, Helen challenges the standard view of how the industry functioned, and his new interpretation, if correct, has wide implications for the role of the upper classes in business; but he cannot be said entirely to have proved his case. The first of Helen's two main conclusions—that *figlinae* means claylands—is clearly right; and he argues convincingly that the *officinator* was often an independent businessman (or woman), moving at will from brickyard to brickyard and even renting claylands from several *domini* simultaneously. But he does not prove that the *officinator*, still defined only in relation to the *dominus* as "the other person in the stamp", was not at other times a paid foreman or manager; and many *domini*, who must have owned the workshops along with the claylands in which they were located, were no doubt more directly concerned with the exploitation of their *figlinae* than Helen is prepared to admit.[14] Thus, despite the considerable advances Helen has made for our understanding of the Roman brick industry, and despite his confronting the interpretive problems that hamper any reconstruction of its organization, the nature of the relationship

[9] See, in addition to the works cited in the preceding note, "*Consules Suffecti* on Roman Brick Stamps", *CP* 39 (1944), pp. 254-255; "Ein datierter Ziegelstempel Theoderichs des Grossen", *RM* 66 (1959), pp. 196-203.

[10] The *Supplement*, first published in *HSCP* 56-57 (1947), pp. 1-128, and the *Indices*, first published in *HSCP* 58-59 (1948), pp. 1-104, have been reprinted as a single volume: *Supplement to Volume XV,1 of the* Corpus Inscriptionum Latinarum *Including Complete Indices to the Roman Brick Stamps* (Cambridge, Mass. 1948 and Rome 1967).

[11] T. Helen, *Organization of Roman Brick Production in the First and Second Centuries A.D.* (Annales Academiae Scientiarum Fennicae, Dissertationes Humanarum Litterarum 5) Helsinki 1975. Separate edition: *AIRF* IX.1 (1977); hereafter cited as Helen, *Organization*. M. Steinby, "La cronologia delle *figlinae* doliari urbane dalla fine dell'età repubblicana fino all'inizio del III secolo", *BullCom* 84 (1974-1975), updated to 1977, pp. 7-132; hereafter cited as Steinby, *Cronologia*. P. Setälä, *Private* Domini *in Roman Brick Stamps of the Empire* (Ann. Ac. Sci. Fenn., Diss. Hum. Litt. 10) Helsinki 1977. Separate edition: *AIRF* IX.2 (1977); hereafter cited as Setälä, *Domini*.

[12] Since Steinby incorporates into her article many of Bloch's conclusions on the dating of particular stamps found *in situ*, consultation of references to her work will often reveal Bloch as the original source. Nevertheless, Steinby's study is cited as the most complete treatment of the subject, and, in order to avoid duplication, references to Bloch's work have for the most part been omitted.

[13] The similarity between Bloch's and Steinby's contributions to the study of brick stamps extends to the dispersion of Steinby's learning throughout numerous publications and to the quantity of new material that she has made accessible, through her "Appendice a *CIL* XV,1" (*BullCom* 86 (1978-1979), pp. 55-88; hereafter, *Appendice*) and her exemplary editions of various collections, for a list of which, see the Bibliography. (I have not seen Steinby's most recent study, *I bolli laterizi dell'Area Sacra di Largo Argentina* in *L'Area Sacra di Largo Argentina I.* Rome 1980).

[14] Cf. *Organization*, pp. 108-109, 130 with the review of V. Righini, *RBPh* 55 (1977), pp. 639-642 and the remarks of R. Friggeri in her review of Setälä's book, *RFIC* 108 (1980), pp. 95-96. As Helen notes (pp. 90, 130), persons appearing only in one-name stamps might have been either landowners or brick producers or, as was quite likely often the case, both. My interpretation of the term *figlinae* is tolerably represented by the translation "brickyards", which I take to mean primarily an area of clay deposits, but also the buildings where the bricks were made and the kilns in which they were fired. In this sense, a fair analogy is found in the English "vineyard", which often implies not only the land under cultivation, but also the winery where the grapes are pressed; in fact, the term most clearly carries this connotation when it appears on the labels of wine bottles as part of a trademark.

between *dominus* and *officinator* in individual cases remains obscure. The term *officinator* still may describe anything from a slave foreman in his master's service to a powerful industrialist of equestrian rank.

Setälä's study of the private *domini* is useful primarily for the mass of prosopographical information that she has gathered about the owners of urban brickyards. Although reviewers have detected various errors in the individual biographies and have questioned some of Setälä's conclusions on the composition of the class of brickyard owners (that it reflects a drastic change in the governing class at the beginning of the second century) and on the transference of land (that it passed mainly by inheritance), her work, if used with care, is nonetheless a valuable tool for students wishing to learn about the families and careers of particular landowners, and it forms a helpful complement to the works of Steinby and Helen.[15]

With the publication in 1977 and 1978 of the first two volumes of the edition of the stamps from Ostia, *Lateres Signati Ostienses*, Steinby and her colleagues set a new standard of precision for editors of brick stamps.[16] The Finns not only equipped each entry in their catalogue with a concise statement of the physical characteristics of the stamp, they also supplemented their verbal descriptions in all but a few instances with photographs of one or more copies. In so doing they enabled students to identify fragmentary copies more securely and provided them with their only sure means of distinguishing stamps that sometimes differ from each other only in measurements, paleography, or disposition of the text. Stamps of this sort were known to Dressel, but Dressel's philological bias led him to concentrate on the text at the expense of the stamp itself; consequently he listed all examples with the same text under the same number in *CIL* XV,1 and signalled any discrepancies in size or orientation only with a general comment at the end of the entry: *sigillum signaculis diversis impressum extat*. Moreover, Dressel was inconsistent in reporting these minor variations, acknowledging them openly at times, recording them incidentally at others, and overlooking them or ignoring them at others still. The systematic identification and description of these variants now seems essential for resolving the chronological problems that they quite often present, for Bloch has pointed out that different stamps with identical text were sometimes used successively over a period of several years.[17] Complete photographic documentation of all the Roman brick stamps is therefore greatly to be desired. True variants can be distinguished from misread copies only through direct visual comparison of two or more examples; and only when legible photographs are readily available can many of the unidentified fragments now scattered throughout the archaeological publications be assigned their proper place in *CIL* XV,1.[18]

The publication in *LSO* of photographs of more than one third of the known urban stamps constituted an important step toward achieving the above mentioned goal; and it marked the arrival of a new era in the study of brick stamps, in which, more than ever before, careful attention to detail has become the first responsibility of those who would edit the material. In publishing the Michigan collection of brick stamps,

---

[15] For corrections of some factual mistakes, see M. Raepsaet-Charlier, *AC* 48 (1979), pp. 772–774; L. Schumacher, *HZ* 229 (1979), pp. 119–122; N. Purcell, *JRS* 71 (1981), pp. 214–215. Cf. also A. M. Small, *Phoenix* 33 (1979), pp. 369–372; H. Halfmann, *Gymnasium* 86 (1979), pp. 208–210; and R. Friggeri, *RFIC* 108 (1980), pp. 93–96.

[16] *Lateres Signati Ostienses* 1. *Testo* by M. Steinby in collaboration with T. Helen (Rome 1978). 2. *Tavole* by M. Steinby (Rome 1977) (*AIRF* VII.1–2); hereafter cited as *LSO*. The edition will eventually comprise four volumes: Volume 3 will contain complete indices to *CIL* XV,1 and its supplements, including an index of *signa* and an alphabetical list of individual lines of text; Volume 4 will consist of a study of the *signa* and forms of stamps and a list of the discoveries *in situ* with relative dating of all the material known to date. For a more complete survey of the scholarship on brick stamps, see Bloch, *BL*, 3–7, and, for the years 1938–1978, Steinby in *LSO*, pp. 11–15 with further bibliography; cf. also pp. 19–21, where mention is made of studies completed by other members of the Finnish team on the imperial ownership of *figlinae* and on their location, on the terminology and paleography of stamps, on stamps of the fourth century, and on the mineralogical composition of ancient bricks. Some of these will be forthcoming in future volumes of *AIRF*.

[17] *AJA* 63 (1959), p. 225 n. 2; see also Steinby's remarks in *MemAccLinc* 17 (1974), pp. 64–65: "bisogna cioè datare i timbri e non i numeri del *CIL*." She notes that the identification of examples impressed from broken or cracked dies is similarly worthwhile since defective copies can usefully be compared with examples made before the die was damaged in order to determine the "internal" chronology of a particular stamp (cf. no. **25**, for example). Occasionally stamps of different persons can be recognized as having come from dies that were carved by the same hand; at the very least this implies an indirect connection between the persons named in the stamps and establishes their contemporaneity (cf. nos. **96, 98**).

[18] For a practical demonstration, see the comment at **83**.

my chief aim has therefore been to make as useful a contribution as possible to a complete and accurate knowledge of the material. If my descriptions of the individual examples and the photographs at the end of the volume should satisfy the needs of future editors, the edition will have well fulfilled its purpose.

II. HISTORY OF THE MICHIGAN COLLECTION

Dates of Acquisition

The collection of some five hundred examples of brick stamps currently located in the Kelsey Museum of Archaeology in Ann Arbor originated in 1893 with Francis W. Kelsey, who included a small fragment of stamped brick among the original purchases he made on behalf of the University of Michigan to found the University's archaeological collections (**139**).[19] An avid antiquarian, Kelsey eventually built the collection to nearly its present size; but in the years immediately following his return to the United States in 1894, Kelsey's duties in Ann Arbor obliged him to participate only indirectly in the wholesale buying of antiquities that the University was then conducting overseas through the agency of his friends and colleagues. One of these was Walter Dennison, who spent the years 1895–1897 in Rome as a Fellow at the newly founded American School of Classical Studies. When Dennison learned on a visit to Pozzuoli in 1897 that the valuable collection of antiquities belonging to the parish priest, Giuseppe De Criscio, was available for purchase, he wrote to Kelsey in search of a donor. Kelsey promptly enlisted the support of a private benefactor and wrote back, encouraging Dennison to further the negotiations that were to result two years later in the shipment to Ann Arbor of 267 of De Criscio's inscriptions, including six new brick stamps and one tile marked with a graffito (**2**; **5**; **6**; **10**; **11**; **85**, Kelsey acc. no. 1060; Appendix 1).

At least one of the examples (**2**), edited by Dressel, had already appeared in *CIL* X, where Mommsen had been permitted to publish nearly two hundred of De Criscio's texts. [20] Four others were included among the sixty-seven inscriptions that Dennison published in 1898 in the *American Journal of Archaeology* (**5**, **6**, **10**, **11**).[21]

In the fall of 1900 Kelsey returned to Italy. There, during the early months and summer of 1901 he purchased 477 examples of urban stamps from dealers and workmen in Rome. Shipped to Ann Arbor and received in the following year, the examples were numbered and many were identified in the summer of 1904 by H. M. Gelston and J. B. Hunter.

Subsequent accessions were small and attracted little interest: a second purchase from De Criscio in 1905 added seven examples to the collection (**3**; **4**; **7**; **8**; **9**; **85**, Kelsey acc. no. 1130; **91**); single copies were obtained from De Criscio's heirs in 1923 (**140**) and from an anonymous donor in 1937. Another eleven arrived with the materials bequeathed to the University by E. B. Van Deman and secured from her heirs in 1938 (**1**, **51**, **133**). Two stamps without text were donated at an uncertain date by a private collector (**134**, **137**); and the acquisition in 1981 of four new stamps (**92**) brought the total number of examples in the Michigan collection to 523, where it stands today.

---

[19] Information regarding the history of the Michigan collection has been gathered from various sources: notes in the Museum records on the epigraphical accessions written by F. W. Kelsey, W. Dennison, O. F. Butler, and J. G. Winter; the *Minutes of the Board of Regents of the University of Michigan* for November, 1899, pp. 458–459; Kelsey's annual diaries (1899–1927), now part of the University of Michigan Historical Collections; O. F. Butler, "Report of the Museum of Classical Archaeology", reprinted from the *Report of the President of the University of Michigan*, 1928–29 (Ann Arbor 1930).

[20] The two copies of *CIL* XV 1146b now in the Kelsey Museum may be the same ones that Mommsen published at *CIL* X 8331, 1: see the comment at **85**. For a list of the 154 texts included in *CIL* X (1883) and the 36 published in its supplement, *EphEp* VIII (1899), that are currently in the Kelsey Museum, see J. H. D'Arms, "Eighteen Unedited Latin Inscriptions from Puteoli and Vicinity", *AJA* 77 (1973), p. 151 nn. 2, 4.

[21] W. Dennison, "Some New Inscriptions from Puteoli, Baiae, Misenum, and Cumae," *AJA* 2 (1898), pp. 373–398. For the 47 inscriptions edited by Dennison that are now in the Kelsey Museum, see J. H. D'Arms, *AJA* 77 (1973), p. 151 n. 5.

Provenance

Little is known about the archaeological context of the examples in the Michigan collection. Kelsey carefully documented the circumstances under which he had acquired the first fragment in 1893: he bought it from an old woman on the road to the Villa Iovis at Capri (**139**). But at the time of his great purchases in 1901, Kelsey made no attempt to record even the general locations where the examples had been discovered, because he had little confidence in the reliability of his informants. Consequently, in regard to the provenance of some 470 examples, we have only this laconic note, written by Kelsey in the Museum accessions catalogue: "They are all from Rome."[22]

Of the fifteen examples from the De Criscio collection, Museum records indicate that nos. **3** and **8** were originally purchased at Cumae; the rest are listed as having been found at Pozzuoli. In a few instances it has been possible to corroborate and supplement this evidence by adducing the independent testimony of Dressel (**2**) and Dennison (**5, 6, 10, 11**), who must have derived their information from De Criscio himself. For the remaining examples we should perhaps be guided by the geographical distribution of the other inscriptions owned by De Criscio, which included the areas around Baiae and Misenum as well as Cumae, although there is no reason to discount Pozzuoli as the actual source of discovery.[23] Only one of the eleven examples from the Van Deman collection bears any indication of provenance:[24] a tag attached to the brick names the legionary camp at Lambaesis (**1**). One stamp without text was evidently discovered near Hadrian's Villa at Tivoli (**134**); and four recent acquisitions, all urban stamps, are said to have come from the Roman villa on Giannutri (**92**, Kelsey acc. nos. 81.1.2–81.1.4).

In not one instance does this meager information provide a secure basis for dating a stamp, nor do the reported places of discovery add anything new to our knowledge of the origin or distribution of ancient bricks.[25]

III. CHARACTER OF THE COLLECTION

The Michigan collection of 523 examples, though small in comparison with the vast inventories of the museums and antiquariums of Rome, is nonetheless the largest known collection of ancient brick stamps preserved outside of Italy. With few exceptions the examples belong to one of two groups—the relatively few Campanian stamps obtained from De Criscio, and the far larger group of urban stamps, most of which were purchased by Kelsey in 1901. The nature of the collection is thus essentially as Kelsey determined it by his methodical acquisitions of that year, methodical because they reveal the care he took to select a representative sampling of the body of Roman brick stamps: the 477 examples collected in Rome yield no fewer than 320 different stamps, which are represented in most cases by a single copy and never by more than nine.

In its high ratio of diverse stamps to number of examples, the Kelsey collection compares favorably with other groups of urban stamps and closely resembles the Di Bagno collection assembled, at least in part, by Gaetano Marini. [26] Like the examples in the Di Bagno collection, the stamps collected by Kelsey show a

---

[22] In another note Kelsey mentioned having bought one example on the Via Nazionale (**98**), another off the Roman Forum (**125**). Where not otherwise indicated in the comments, all examples included in the Catalogue are part of the acquisitions at Rome in 1901. In the same year Kelsey travelled to Greece, where he purchased a Greek brick stamp at Athens (**138**) and a fragment of brick inscribed with a graffito at Argos (Appendix 2).

[23] For the distribution of the inscriptions in the De Criscio collection, see Dennison, *AJA* 2 (1898), pp. 373–398 and *CIL* X, *passim* (with D'Arms, *AJA* 77 (1973), pp. 151 nn. 2, 4).

[24] Van Deman's main interest was in determining the construction date of architectural monuments; for this she considered brick stamps of little value as evidence, even when found *in situ* (cf. Bloch, *BL*, p. 12).

[25] But cf. nos. **1** (dating); **16** (origin).

[26] A total of 366 different stamps are represented by the 523 examples that constitute the entire Michigan collection. For Marini and the Di Bagno collection (192 stamps, 302 examples), see V. Righini, *I bolli laterizi romani. La collezione Di Bagno.* Bologna 1975, pp. 6–10. Other collections: Ostia: 1378 stamps, 9376 examples in 1971 (*LSO*); Antiquarium Comunale: 900 stamps, 4650 examples in 1937 (Bloch); Antiquariums of the Forum and Palatine: 770 stamps, 2700 examples in 1973 (Steinby); Torre Angela: 307 stamps,

relatively balanced distribution over the categories established by Dressel: 1 of the 8 public stamps is represented; 34 of the 73 *figlinae* reported in *CIL* XV,1,[27] or approximately half, are attested in the Kelsey examples, as are approximately one quarter of the imperial stamps (20 out of 83), one seventh of the private stamps (105 out of 764), and one third of the *officinae* of the age of Diocletian (6 out of 21). And like the compilers of the Di Bagno collection, Kelsey showed a penchant for stamps with a consular date: of the 399 known varieties, the Michigan collection includes 70, which represent nearly half (23) of the 49 years between 110 and 164 that are attested in brick stamps. More important, the collection illustrates the duration of the Roman brick industry, spanning a period from the end of the Republic to the beginning of the sixth century A.D. It thus reflects Kelsey's aim of gathering a small body of material typical of the whole, to serve as an introduction to the study of brick stamps for students in the United States.

But the Michigan collection also contains material of particular interest to specialists: 13 new stamps; 5 fragments of unedited stamps; 10 examples that complete uncertain readings in *CIL* or the *Supplement*; and 12 examples that constitute corrections or variants of previously published stamps. Detailed commentary on the individual examples appears below at the appropriate entries in the Catalogue, but it may be useful to mention here the more noteworthy items of interest to students of the Roman brick industry.

One previously unpublished fragment adds a new member, [Am]aranth[us], to the rare *gens Marsidia* (**94**); another evidently records the name of a slave, Ant[imachus], who is known to us already as a freedman, Aelius Antimachus, and associates him with a bust of Mercury (**124**). From no. **107** we learn that a slave Anthus, at one time the property of Marcus and Titus Varienus, was later owned by only Marcus, whose activity in the brick industry of the first century A.D. is well documented by stamps of his slaves and freedmen discovered at Pompeii and Rome. Other new figures involved in brick manufacturing are the Campanians St. Fl(avius) ? M( ) (**10**) and the slave Serapio (**9**), and, from Rome, a C. Umbrius ? S[ym]phorus (**110**), a Ser( ) Rub( ) (**128**), and two persons known only by their initials A( ) V( ) (**69**) and O( ) V( ) (**99**) (cf. also nos. **125–127**).

A stamp of the *figlinae Caelianae* previously known only from a single copy discovered in Africa is now attested also at Rome (**16**). One example preserving a carver's error in the spelling BIPIDALE for BIPEDALE apparently came from the same die as a copy published in *LSO*, where the mistake has been corrected by recutting the die (**32**; cf. **89**, **31**). In at least one other stamp the carver mistakenly inserted a mark of punctuation in the middle of a word (**70**; cf. **123**). Finally, a complete example of a stamp that Dressel did not know and that Bloch did not see shows a new decoration pertaining, it seems, to the cult of Isis: *quindecim semina* (?) (**86**).

IV. THE EDITION

Order of Entries in the Catalogue

The Catalogue includes all stamps in the Kelsey Museum that have not previously been photographed and those which, having been photographed elsewhere in fragmentary or imperfect condition, are better preserved in the Michigan examples.[28] The selection comprises, in addition to the 18 unedited stamps and

---

533 examples in 1971 (Coste); Santa Maria Maggiore: 159 stamps, 407 examples in 1973 (Steinby); Vatican necropolis: 55 stamps, 69 examples in 1971 (Steinby). Well over half of the 7000 examples preserved in the Museo Nazionale Romano have by now been identified by L. Camilli and F. Taglietti, who are preparing an edition of the entire collection. See the Bibliography for references.

[27] The number of recognized *figlinae* has grown considerably since Dressel's day, not only from the addition of a few names attested on subsequently discovered stamps, but more significantly from the realization that many *figlinae* were divided into subsections (as Dressel had already noted in the cases of the *figlinae Domitianae* and *Oceanae*). Thus Bloch counted 85 *figlinae* in 1947 (*Supplement*, pp. 125–127), Steinby 94 in 1977 (*Cronologia*, pp. 110–111). By these figures the number of *figlinae* attested in the Michigan examples would be correspondingly higher.

[28] It has not been possible to photograph and include in the Catalogue every stamp in the Michigan collection. Consequently, in Concordance II, where every example in the collection is accounted for, copies that can be identified with numbers in *CIL* are

the 21 completions, corrections, and variants mentioned above (p. 7), 65 copies of stamps published in *CIL*, 23 examples that differ from the versions in *CIL* only in the addition of one or more marks of punctuation, 9 stamps without text, and 4 modern or foreign stamps. Three bricks inscribed before firing with graffiti are included in an appendix. In accordance with the procedure established by Bloch in the *Supplement*, the stamps are presented in the order in which they appear in *CIL*, although, as has been noted above (p. 2), Dressel's arrangement is in various ways antiquated.[29] New stamps are inserted at the appropriate place when the text is complete enough to allow the stamp to be located according to Dressel's system; fragments that cannot be located with certainty are assembled after the stamps from *CIL* (**124–128**). These are followed by the stamps without text (**129–137**) and the modern or foreign stamps (**138–141**).

Composition of the Individual Entries

The format used in presenting the texts is, with slight modifications, essentially that adopted in *LSO*. Each entry in the Catalogue consists of a description of the physical characteristics of the stamp, dimensions of the brick, the Kelsey Museum accession number or numbers of the examples in the Michigan collection, reference to *CIL*, an indication of the form of the stamp, a transcription and interpretation of the text, comments, and a photograph in the Plates at the end of the volume. The omission of descriptions of the clay perhaps requires some justification. In previous publications the information supplied by scholars expert in everything but petrology and mineralogy has seemed to offer little promise of determining the ultimate origin of the brick. Although T. Darvill has recently achieved significant results by pursuing this line of research, his conclusion that "the tile industry [of Roman Britain] seems to be considerably more complex than was previously thought" suggests that fabric analysis of bricks produced in other parts of the empire should likewise be conducted by specialists in sedimentology and spectroscopic techniques.[30] My own untrained observations have therefore seemed inadequate and have consequently been suppressed.

Dimensions of the stamp are given in centimeters, as are all measurements: for rectangular stamps with letters in relief, in the order length, then height. The sign + follows the greatest measurement of a fragmentary example; the sign ± indicates that the measurement has been estimated on the basis of two or more fragments. For rectangular stamps with impressed letters (*litteris cavis*) the measurements are normally based on the length of the text and the height of the letters, since the light application of the die often leaves no trace of the border of the stamp; otherwise the abbreviation *marg. vid.* (for *margines videntur*) after the numbers indicates that the margins have been included in the measurement.

For circular and orbicular stamps the sign ± indicates that the diameter has been calculated by doubling the radius; whenever applicable, the diameter of the orbiculus follows that of the stamp. The measurements of all stamps are subject to a certain margin of error (never more than 5 mm.) attributable to the variable dryness of the clay when the stamp was impressed, the sliding of the die along the surface of the brick, and various other factors (see further *LSO*, p. 25).

The heights of the letters are given in the same order in which the lines of text are transcribed, that is, from the uppermost line for rectangular stamps and from the outermost line, in most cases, for circular and

provided, wherever possible, with references to a source where a photograph or drawing of the stamp can be found. There remain eleven examples that are so badly worn as to be barely legible; these too have been excluded from the Catalogue.

[29] Occasionally this has produced a notable incongruity, as, for example, in the case of nos. **7** (= *CIL* X 8042, 66) and **91** (= *CIL* XV 1237): the stamps are closely related to each other and are evidently of Campanian origin—indeed, both examples in the Kelsey Museum are from the De Criscio collection—yet one appears among the stamps of Campania, the other in the *lateres privati* section of *CIL* XV,1. In this instance, disregard of Dressel's practice of classifying stamps by source of discovery has resulted in the same wrong placement of *CIL* XV 1237 among the stamps of Rome. For different aspects of the same issue, cf. nos. **8** and **85**.

[30] T. Darvill, "A Petrological Study of LHS and TPF Stamped Tiles from the Cotswold Region," *BAR IntSer* 68 (1979), pp. 309–349. See also the remarks of V. Righini in *Epigraphica* 38 (1976), p. 192. In the Finnish team's study of the stamps from Ostia, analysis of the physical characteristics of the bricks was entrusted to a mineralogist, Hannu Appelqvist (cf. *LSO*, p. 20).

semicircular stamps. The minimum and maximum letter height is recorded for each line; letters which exceed the median range are enclosed in parentheses.

Auxiliary lines are the narrow lines that delineate the lines of text; they are reported in the same order as the heights of the letters. A dash is used to mark the certain absence of auxiliary lines, a question mark to indicate that the reading is uncertain.

Dimensions of the brick are given in the following order: maximum breadth across the face, height, depth. In the few cases in which the entire brick is preserved, its type is indicated by the abbreviation *bip.* for *bipedalis*, *sesq.* for *sesquipedalis*, or *teg.* for *tegula* (in the strict sense meaning "roof tile", not, as the term is used in the stamps themselves, in reference to any large brick); otherwise the measurements are preceded by the abbreviation *fr.* for *fragmentum* and are followed, individually, by the sign + to indicate that the measurement is somewhat less than the original.

The Kelsey Museum accession number or numbers of the example or examples in the Michigan collection is recorded in parentheses after the list of measurements. If a stamp is represented by more than one copy, the measurements and the photograph are of the first example listed, whereas the number of auxiliary lines and, in the transcription of the text, the marks of punctuation have in a few cases been incorporated from the other examples.

The reference to *CIL* is followed by a question mark if the identification is uncertain. Variations in respect to *CIL* are defined as follows: A completion (*completum*) is any example that improves or confirms a version given in *CIL* as imperfect or uncertain; the simple addition of one or more normal triangular marks of punctuation, however, is signalled only by an asterisk (*). Any change of a transcription given as complete and certain constitutes a correction (*correctum*). A variant (*variatio*) is any stamp that differs from a previous publication in ligatures, abbreviations, or ornament, whereas differences in form, formula, *signum*, or content—including even the addition or substitution of a single letter—are sufficient criteria for considering a stamp to be new (*novum*).

Reference has been made above (p. 4) to stamps with identical text which nonetheless differ graphically and which Dressel acknowledged with the comment *sigillum signaculis diversis impressum extat*. The only deviations allowed are in measurements, punctuation, number of auxiliary lines, shape of the letters, and disposition of the text. These stamps are marked in the Catalogue with Roman numerals after the reference to *CIL*. As a preliminary step to the necessary revision of *CIL* XV,1, whenever a Michigan example presents a new variant of this sort, I have tried to mention in the comments all previously reported variants and, after assigning to each a number based chronologically on the date of publication, to include the new stamp in the system by adding it to the end of the list.[31] Thus, for example, no. **47** is identified as *CIL* XV 525c III, although there is only one version of *CIL* XV 525c in the Michigan collection. The same procedure has been followed with multiple variants of the same stamp (*variationes*), but the different variants are designated with Arabic numerals to distinguish them as a class from the stamps that differ from each other only graphically (cf. **13**, for example).

The form of the stamp is indicated by one of nine symbols:

     ☐   rectangular form, letters in relief. No symbol is used for rectangular stamps with impressed letters, or for letters in relief without margins.

     ☋   semicircular form with orbiculus.

     ○   circular form without orbiculus, center flush.

     ◎   circular form without orbiculus, center in relief.

---

[31] The single exception to this rule is no. **113**, where the authority of *LSO* has been preferred to the priority of the publication of the collections of the Forum and Palatine Antiquariums.

☽	lunate form; the orbiculus extends beyond the center of the stamp.
☽	orbicular form with the center in relief.
☽	orbicular form; the orbiculus has the height of one line of text.
☽	orbicular form not included in one of the preceding categories.
◯	octagonal form.

The stamps without text are described verbally.

The transcription of the text necessarily falls short of the ideal established by Dressel, who aimed at reproducing in print an image of the stamp faithful to the original, but it nonetheless preserves some of the characteristic features of brick stamps: the lines of circular and semicircular stamps have been centered as accurately as possible, as have those of rectangular stamps where the text is in reality centered (it has not been possible to block up rectangular stamps); texts in mirror writing, reversed or inverted letters, and A's without the crossbar are reproduced typographically; palm fronds, arrows, stars, and ivy are represented by symbols.

Ligatures are rendered by the sign ‿ below the text. Fragmentary letters are marked by a dot below the letter if the reading is uncertain or by a dot alone if no letter can be suggested. An oblique slash (/) indicates that a letter has been erased or is missing because of a defect in the die. For circular stamps, straight lines of text and lines that gradually increase in height towards the center are noted with the formulae adopted by Dressel: *linea (fere) recta; vers. falc.* (for *versibus falcatis*). *Signa* and decorative elements that cannot be graphically reproduced are described in Latin.

The lines of text are normally printed in the order top to bottom for rectangular stamps and outermost to innermost for circular and semicircular stamps. Stamps of the fourth century, though circular in form, are treated as rectangular stamps since the text is impressed in straight lines. Orbicular stamps *versibus falcatis* are likewise transcribed as rectangular stamps when the reading clearly begins with the uppermost line (cf. **76, 82**); the regular order is maintained, however, when the lines of text are concentric (cf. **62**).

In **the interpretation of the text** the following editorial conventions are used:

[ ]	integrated text.
( )	interpretation.
[[ ]]	erased text or text missing because of a defect in the die.
< >	text mistakenly omitted by the carver and added in the interpretation.
⌐ ¬	corrected text.
{ }	text incised by mistake and eliminated in the interpretation.
?	interpretation uncertain.

In keeping with a practice first introduced by Steinby in *LSO*, commas have been eliminated from the interpretations and terms of clarification have been kept to a minimum.[32] Virtually the only exceptions are the words used in connection with the names of *figlinae* in reference to the clay products themselves. These

---

[32] The omission of commas from the interpretation of brick-stamp texts—in Steinby's words, "una piccola rivoluzione nella tradizione dell'epigrafia doliare"—must in the future become standard procedure. Since commas serve mainly to define the role in the brick industry of persons appearing in the text, their usefulness is essentially limited to the rare cases in which a person's status is ambiguous from the text in question but is known from other sources; even in these cases the problem is more clearly resolved through explanatory terms. For most stamps the use of commas in the interpretation is either superfluous or misleading. See further Steinby in *LSO*, pp. 29–30, and, for the problems involved in the reading of stamps, Helen, *Organization*, pp. 31–35.

explain the gender of the proper name and in second-century stamps are determined by the usage in other stamps from the same brickyard; for first-century stamps the term adopted is generally *tegula.* Whenever these explanatory terms are added to the interpretation, they are signalled by the abbreviation *sc.* to indicate that specification was not considered necessary by the person who made the stamp. Thus, for example, one finds *Brut*(*iana* sc. *tegula*) (**13**), but *Sulp*(*icianum* sc. *opus*) (**51**).

The only abbreviations not expanded in the interpretations are those considered to be so familiar as to need no explanation, specifically the various *praenomina* and the abbreviation *cos.* (for *consulibus*) when it occurs in the consular-dating formula. Forms such as *ficlinae, feglinae, peredis,* and the like have been allowed to stand as representing genuine linguistic variants, whereas the comment "sic" in the margin has been reserved for true and idiosyncratic errors on the part of the person who made the stamp.

In **the comments** I have not always followed the principle adopted in *LSO* of not repeating information already supplied by previous editors. Rather, I have tried to make the stamps more accessible to students unfamiliar with the material by providing brief remarks on the dating of each stamp and on the persons mentioned in the text, and by supplementing these with frequent bibliographical references to more complete discussions in the works of Bloch, Steinby, and the other members of the Finnish team.[33] Furthermore, whenever in the interpretation of a particular example I have touched on an issue of broader significance concerning brick stamps, I have taken the opportunity to refer the reader to the principal literature on the subject.[34] Of greater use to specialists than these general comments will be the descriptions of individual examples and the references to preceding publications of identical and closely related stamps.

Wherever possible, I have tried to distinguish between the terms "stamp", "die", and "example": a "stamp" is any impressed design that exists (or could exist) in multiple copies, whereas an "example" is any particular copy of a stamp; the "die" is the instrument used to imprint the design on the surface of the brick. In many contexts the difference in meaning between "stamp" and "example" is negligible, but the distinction becomes important when "example" refers to a copy that differs from the "ideal" in particulars that can be attributed to an imperfect state of preservation, either of the brick or of the die. Strictly speaking, an example impressed from a broken or dirty die constitutes a new stamp in that it could exist in numerous copies, but in order to avoid confusion a stamp is considered to be different only if it has been impressed from a different die or one that has been intentionally altered (see also above, p. 9).

The provenance of the few examples for which Museum records provide specific information is reported accordingly. Examples from the De Criscio collection are identified as such in the comments. Where not otherwise indicated, all examples are part of Kelsey's acquisitions in 1901 and were purchased in Rome (see above, p. 6).

For consular-dated stamps, the year is given in the margin preceded by the abbreviation *a.* for *anno.* Otherwise, approximate dating of each stamp is based on the following criteria, which are listed in descending order of usefulness: other examples found *in situ*; mention in the text of persons known from other sources; form of the stamp and syntactical composition of the text; and, especially for early stamps, paleographical considerations.[35]

The stamps without text (*sine textu*) are assembled after the fragments of unedited stamps. Because of the uncertainty inherent in identifying stamps without text merely by reference to preceding editions, all nontextual stamps in the Michigan collection are published in the Catalogue, including those that have previously been photographed. Unfortunately it has not been possible to reproduce the stamps photo-

---

[33] In a few instances observations attributed to Steinby appear in the comments without citation; these are her unpublished opinions, communicated to me through correspondence and notes on a preliminary version of the manuscript, for both of which I am extremely grateful.

[34] See, for example, no. **13** on the consular-dating of brick stamps, no. **36** on the use of *signa* as distinguishing emblems, no. **70** on the mutual exclusiveness of the groups of *officinatores* and *domini*, no. **85** on the geographical distribution of urban stamps, no. **107** on *societas* in the brick industry, no. **111** on stamps of the age of Diocletian, and Appendix 1 on the duration of the brickmaking season and on the average number of bricks produced by a workman in a single day.

[35] For the relative merits of the various means of dating stamps, see Steinby, *Cronologia*, pp. 16–22. For a more complete explanation of the procedures adopted in editing the Michigan examples, see Steinby in *LSO*, pp. 25–31.

graphically in their original size; consequently extra care has been taken in describing the forms and recording the measurements.

The modern and foreign stamps are collected at the end of the Catalogue. In evaluating the authenticity of nos. **139** and **140** I have been guided particularly by the judgement of Professor Steinby, who examined the photographs.

Three graffiti inscribed on bricks are presented in an appendix, since they are relevant to the study of ancient brick production if not, strictly, to the study of brick stamps.

The Concordances and Indices are designed mainly to serve as guides to the Catalogue, but Concordance II contains also material not found elsewhere in the edition. The first concordance is composed of the variations in respect to *CIL*: the unedited stamps, the completions, corrections, and variants, and the stamps identified with the same number in *CIL*. Concordance II comprises all examples in the Michigan collection: the examples that can be identified with stamps in *CIL* or the *Supplement* are listed first, then the examples identifiable with numbers in the *Supplement* are listed separately; these are followed by the unedited fragments, the stamps without text, the modern and foreign stamps, and the unidentified fragments. Wherever possible, a source where a photograph or drawing of the stamp can be found is cited as a control on the identification of the Michigan example.

The indices of *nomina, cognomina,* emperors, consuls, brickyards, and *notabilia varia* are elaborated after the model of Bloch's *Indices*; an index of *signa*, which includes the decorative elements found on stamps, has been added after the index of *figlinae*. In the interest of those who may use this edition in conjunction with Bloch's *Indices*, material not contained in the latter is printed in boldface type.[36]

In **the Plates** a photograph of each stamp in the Catalogue is published in the order in which it occurs in the text; consequently reference to the Plates has been omitted from the individual entries. In the captions the Catalogue number, printed in boldface type, is followed first by the corresponding reference to *CIL* or the *Supplement* and then, enclosed in parentheses, the Kelsey Museum accession number of the example photographed.

---

[36] The entries printed in boldface include not only information contained in unedited urban stamps and new interpretations of material already known to Bloch, but also material pertaining to the stamps from Campania and Africa, which has no place in the indices to *CIL* XV, and, in the section of the *notabilia varia* entitled "Variants of individual letters", a few items that seem to have been omitted from Bloch's *Indices* because of an oversight. The index of *signa* is entirely new and is therefore printed in ordinary type. The modern or foreign stamps and the graffiti have been altogether excluded from the indices.

BIBLIOGRAPHY AND ABBREVIATIONS

Below is a list of the authors and works cited in the comments. The bibliography does not include articles in standard works of reference, except for Steinby's valuable survey in the *RE*, or the reports of archaeological discoveries published regularly in the *Notizie degli Scavi* and the early volumes of the *Bullettino Comunale*. Works cited only in the Introduction have also been omitted.

AE = *L'Année Épigraphique.*
AEA = *Archivo Español de Arqueologia.*
AIRF = *Acta Instituti Romani Finlandiae.*
AJA = *American Journal of Archaeology.*
Ashby, T., "The Classical Topography of the Roman Campagna, Part III (The Via Latina), Section II", *PBSR* 5 (1910), 213–432.
BAR IntSer = *British Archaeological Reports. International Series.*
Blake I = M. E. Blake, *Ancient Roman Construction in Italy from the Prehistoric Period to Augustus.* Washington 1947.
Blake II = M. E. Blake, *Roman Construction in Italy from Tiberius through the Flavians.* Washington 1959.
Blake-Bishop III = M. E. Blake and D. Taylor Bishop, *Roman Construction in Italy from Nerva through the Antonines.* Philadelphia 1973.
Bloch, *BL* = H. Bloch, *I bolli laterizi e la storia edilizia romana.* Reprinted, with indices, from *BullCom* 64 (1936), 141–225; 65 (1937), 83–187; 66 (1938), 61–221. Rome 1947 and 1968.
Bloch, H., "*Consules Suffecti* on Roman Brick Stamps", *CP* 39 (1944), 254–255.
Bloch, *Suppl.* = H. Bloch, "The Roman Brick-Stamps not Published in Volume XV,1 of the *Corpus Inscriptionum Latinarum*", *HSCP* 56–57 (1947), 1–128. See Bloch, *Indices* below.
Bloch, *Indices* = H. Bloch, "Indices to the Roman Brick-Stamps Published in Volumes XV,1 of the *Corpus Inscriptionum Latinarum* and LVI–LVII of the *Harvard Studies in Classical Philology*", *HSCP* 58–59 (1948), 1–104. The *Supplement* and *Indices* have been reprinted as a single volume: *Supplement to Volume* XV,1 *of the* Corpus Inscriptionum Latinarum *Including Complete Indices to the Roman Brick-Stamps.* Cambridge, Mass. 1948 and Rome 1967.
Bloch, *Ostia* I = H. Bloch, "I bolli laterizi e la storia edilizia di Ostia", Appendix in *Scavi di Ostia* I. *Topografia generale,* ed. G. Calza, G. Becatti, I. Gismondi, G. De Angelis D'Ossat, H. Bloch. Rome 1953, 215–227.
Bloch, H., "Sette Bassi Revisited", *HSCP* 63 (1958), 401–414.
Bloch, H., "Ein datierter Ziegelstempel Theoderichs des Grossen", *RM* 66 (1959), 196–203.
Bloch, H., "The Serapeum of Ostia and the Brick-Stamps of 123 A.D.", *AJA* 63 (1959), 225–240.
BullCom = *Bullettino della Commissione Archeologica Comunale di Roma.*
Cagnat, R., *L'Armée romaine d'Afrique et l'occupation militaire de l'Afrique sous les empereurs.* Paris 1892.
Camilli, L., "Contributo allo studio dei bolli laterizi del Museo Nazionale Romano", *RendAccLinc* 28 (1973), 298–312, 326–348.
Camilli, L., "Nuovo contributo allo studio dei bolli laterizi del Museo Nazionale Romano", *RendAccLinc* 34 (1979), 189–191, 194–196, 204–212.
Castrén, *Ordo* = P. Castrén, *Ordo Populusque Pompeianus: Polity and Society in Roman Pompeii.* (*AIRF* VIII). Rome 1975.
CIL = *Corpus Inscriptionum Latinarum.*
CIL XV *A* = Steinby's Appendix: see Steinby, *Appendice*. The abbreviation *A* is used to indicate numbers in Steinby's catalogue, *Appendice* in reference to her comments in the text.
CIL XV *S* = Bloch's *Supplement*: see Bloch, *Suppl.* The abbreviation *S* is used to indicate numbers in Bloch's catalogue, *Suppl.* in reference to his comments in the text.

Coste, J., "Ricerca dei bolli laterizi in una zona dell'Agro romano, Torre Angela", *RendPontAcc* 43 (1970–1971), 71–108.

Cozzo, G., *Una industria nella Roma imperiale. La corporazione dei figuli ed i bolli doliari* (*MemAccLinc* ser. VI, vol. V, 4). Rome 1936.

CP = Classical Philology.

D'Arms, *CSS* = J. H. D'Arms, *Commerce and Social Standing in Ancient Rome.* Cambridge, Mass. 1981.

Dennison, W., "Some New Inscriptions from Puteoli, Baiae, Misenum and Cumae", *AJA* 2 (1898), 373–398.

De Rossi, *Tellenae* = G. M. De Rossi, *Forma Italiae* I.4. *Tellenae.* Rome 1967.

Descemet, Ch., *Inscriptions doliaires latines* (Bibl. des écoles franç. d'Athènes et de Rome 15). Paris 1880.

Dressel, H., *Corpus Inscriptionum Latinarum* XV,1. Berlin 1891.

Garofalo Zappa, *MiscGR* III = G. Garofalo Zappa, "Nuovi bolli laterizi di Ostia", *Terza miscellanea Greca e Romana.* Rome 1971, 257–289.

Gatti, *Navi* = Guglielmo Gatti, "I bolli laterizi delle navi" in G. Ucelli, *Le navi di Nemi.* Rome 1950, 337–348.

Gianfrotta, *Castrum Novum* = P. A. Gianfrotta, *Forma Italiae* VII.3. *Castrum Novum.* Rome 1972.

Helen, *Organization* = T. Helen, *Organization of Roman Brick Production in the First and Second Centuries A.D. An Interpretation of Roman Brick Stamps* (Annales Academiae Scientiarum Fennicae, Dissertationes Humanarum Litterarum 5). Helsinki 1975. Separate edition: *AIRF* IX.1 (1977).

Helen, T., "A Problem in Roman Brick Stamps: Who Were Lucilla N(ostra) and Aurel(ius) Caes(ar) N(oster), the Owners of the *Figlinae Fulvianae* ?", *Arctos* 10 (1976), 27–36.

Helen, T., "The Non-Latin and Non-Greek Personal Names in Roman Brick Stamps and Some Considerations on Semitic Influences on the Roman Cognomen System", *Arctos* 15 (1981), 13–21.

HSCP = Harvard Studies in Classical Philology.

JRS = Journal of Roman Studies.

Kajanto, *LC* = I. Kajanto, *The Latin Cognomina* (Societas Scientiarum Fennica, Commentationes Humanarum Litterarum 36.2). Helsinki 1965.

LSO = Lateres Signati Ostienses 1. *Testo* by M. Steinby in collaboration with T. Helen. Rome 1978. 2. *Tavole* by M. Steinby. Rome 1977. (*AIRF* VII.1-2).

Lugli, G., *La tecnica edilizia romana* I-II. Rome 1957.

Marini, G., *Iscrizioni antiche doliari.* Rome 1884.

MemAccLinc = Atti della Accademia Nazionale dei Lincei. Memorie.

MemPontAcc = Atti della Pontificia Accademia Romana di Archeologia. Memorie.

Mingazzini, P., "Elenco di bolli di mattoni pubblici", *RendAccLinc* 25 (1970), 403–429.

Monaco, E., "Laterizi bollati dalla *Domus Tiberiana*", *RendPontAcc* 48 (1975-1976), 309–313.

NSc = Notizie degli Scavi di Antichità.

PBSR = Papers of the British School at Rome.

PIR[1] = Prosopographia Imperii Romani Saec. I, II, III, ed. E. Klebs, H. Dessau, P. von Rohden. Berlin 1896–1898.

PIR[2] = Prosopographia Imperii Romani Saec. I, II, III, ed. E. Groag, A. Stein, L. Petersen. Berlin—Leipzig 1933– .

Purcell, N., Review of Helen, *Organization* and Setälä, *Domini, JRS* 71 (1981), 214–215.

Quilici, *Collatia* = L. Quilici, *Forma Italiae* I.10. *Collatia.* Rome 1974.

RE = Paulys Realencyclopädie der classischen Altertumswissenschaft, ed. G. Wissowa, W. Kroll, K. Mittelhaus, K. Ziegler. Stuttgart 1894–1978.

RendAccLinc = Atti della Accademia Nazionale dei Lincei. Rendiconti.

RendPontAcc = Atti della Pontificia Accademia Romana di Archeologia. Rendiconti.

Righini, V., *I bolli laterizi romani. La collezione Di Bagno.* Bologna 1975.

RM = Mitteilungen des Deutschen Archäologischen Instituts. Römische Abteilung.

Rodríguez-Almeida, E., "Sellos de ladrillos encontrados en Gabii, en las Campañas 1962 y 1965", *Cuadernos de Trabajos de la Escuela Española de Historia y Arqueología en Roma* 12 (1969), 47–59.

S = Bloch's *Supplement:* see *CIL* XV *S.*

Schulze, *LE* = W. Schulze, *Zur Geschichte lateinischer Eigennamen* (Abhandlungen der königlichen Gesellschaft der Wissenschaften zu Göttingen, Philologisch-Historische Klasse, Neue Folge V.5). Berlin 1904; reprinted 1966.

Setälä, *Domini* = P. Setälä, *Private Domini in Roman Brick Stamps of the Empire. A Historical and Prosopographical Study of Landowners in the District of Rome* (Annales Academiae Scientiarum Fennicae, Dissertationes Humanarum Litterarum 10). Helsinki 1977. Separate edition: *AIRF* IX.2 (1977).

Smith, A. C. G., "The Date of the 'Grande Terme' of Hadrian's Villa at Tivoli", *PBSR* 33 (1978), 73–93.

Solin, *Beiträge* = H. Solin, *Beiträge zur Kenntnis der griechischen Personennamen in Rom* (Societas Scientiarum Fennica, Commentationes Humanarum Litterarum 48). Helsinki 1971.

SPASR I = G. J. Pfeiffer, A. W. Van Buren, H. H. Armstrong, "Stamps on Bricks and Tiles from the Aurelian Wall at Rome", *Supplementary Papers of the American School of Classical Studies in Rome* I (1905), 1–86.

Steinby, *AIRF* VI = M. Steinby, "I bolli laterizi" in *Le iscrizione della necropoli dell'autoparco vaticano,* published under the direction of V. Väänänen (*AIRF* VI). Rome 1973, 171–204.

Steinby, *Battistero* = M. Steinby, "I bolli laterizi", Appendix in G. Pelliccioni, *Le nuove scoperte sulle origini del Battistero Lateranense* (*MemPontAcc* XII.1). Rome 1973, 115–125.

Steinby, M., "Le tegole antiche di Santa Maria Maggiore", *RendPontAcc* 46 (1973–1974), 101–133.

Steinby, M., "I bolli laterizi degli Antiquari del Foro e del Palatino", *MemAccLinc* 17 (1974), 59–109.

Steinby, *Cronologia* = M. Steinby, *La cronologia delle* figlinae *doliari urbane dalla fine dell'età repubblicana fino all'inizio del III secolo.* Rome 1976. Updated to 1977, *BullCom* 84 (1974–1975), 7–132.

Steinby, M., "Ziegelstempel von Rom und Umgebung", *RE* Supplb. XV (1978), 1489–1531.

Steinby, *Pompei* = M. Steinby, "La produzione laterizia" in *Pompei 79,* ed. F. Zevi. Naples 1979, 265–271.

Steinby, *Appendice* = M. Steinby, "Appendice a *CIL.* XV, 1", *BullCom* 86 (1978–1979), 55–88.

Taglietti, F., "Contributo allo studio dei bolli laterizi del Museo Nazionale Romano", *RendAccLinc* 28 (1973), 313–323, 326–348.

Taglietti, F., "Nuovo contributo allo studio dei bolli laterizi del Museo Nazionale Romano", *RendAccLinc* 34 (1979), 191–194, 196–204.

Tomlin, R., "Graffiti on Roman Bricks and Tiles Found in Britain", in *Roman Brick and Tile,* ed. A. McWhirr (*BAR IntSer* 68). Oxford 1979, 231–252.

Van Essen, *Santa Prisca* = C. C. Van Essen, "Inventory of Brickstamps" in M. J. Vermaseren and C. C. Van Essen, *The Excavations in the Mithraeum of the Church of Santa Prisca in Rome.* Leiden 1965, 243–337.

Veny, C., "Algunas marcas de ladrillos y tejas Romaños encontrados en Mallorca", *AEA* 39 (1966), 156–166.

Weaver, *FC* = P. R. C. Weaver, *Familia Caesaris: A Social Study of the Emperor's Freedmen and Slaves.* Cambridge 1972.

ABBREVIATIONS AND EDITORIAL CONVENTIONS

N.B. For the symbols used to represent the shapes of the stamps and the critical signs used in the interpretations of the texts, see the Introduction, pp. 9–10.

> a. = anno
> bip. = bipedalis
> compl. = completum
> corr. = correctum
> d. = dexter, dextra
> ds. = dextrorsum
> fr. = fragmentum
> lin. = lineae auxiliares
> litt. = litterae
> marg. vid. = margines videntur
> N. = novum
> orb. = orbiculus
> sesq. = sesquipedalis
> sig. = sigillum
> s. = sinister, sinistra
> ss. = sinistrorsum
> teg. = tegula
> v. = versus
> var. = variatio
> vers. falc. = versibus falcatis
> (*) = mark(s) of punctuation added with respect to *CIL*

CATALOGUE

1.   Sig. 9.6+, 3.5; litt. 2.1–2.2; fr. 17+, 13+, 3.8. – (29541).

CIL VIII 10474, 1 var. (*)   ▭   ·LEG·III̅·AV̬[

   Leg(ionis) (tertiae) Au[g(ustae) ?].

Cf. *CIL* VIII 22631, 2. The stamp was given to the Kelsey Museum by E. B. Van Deman, who recorded its provenance as Castra Lambaesitana, where the great majority of the stamps of the Third Augustan legion have been found (for an extensive catalogue, with illustrations, of the legionary stamps found in North Africa, see R. Cagnat, *L'Armée romaine d'Afrique*, 433–448). The legion was moved to Lambaesis possibly before the death of Trajan (Cagnat, *op. cit.*, 501), and the permanent camp had just been built when Hadrian visited the site in A.D. 128 (cf. Wilmann's comment at *CIL* VIII 2532). The stamp can perhaps be dated to the intervening period.

2.   Sig. 7.0, 2.6–2.9; litt. 1.8–2.0; fr. 14+, 13+, 2.8. – (1064).

CIL X 8042, 19b   ▭   M A̬RRI

   M. Arri.

This example from the De Criscio collection (see the Introduction, p. 6) is probably the same copy that Dressel described in *CIL* X: see the comment *ad loc*. The stamps of M. Arrius, dated by Steinby to no later than the Augustan or Tiberian age, are frequent in Pompeii and are perhaps to be associated with the *figlinae Arrianae* known from a waxed tablet discovered at Herculaneum to have been the property of Poppaea Augusta in A.D. 63 (see Steinby, *Pompei*, 267, 271; for the Pompeian Arrii, see also Castrén, *Ordo*, p. 137 no. 42). For the problems involved in determining the role in the brickmaking industry of persons appearing only in one-name stamps, see Helen, *Organization*, 89–96. The stamp has not been photographed previously.

3.   Sig. 4.5, 1.9 (marg. vid.); litt. 0.7, 0.7; fr. 39+, 28+, 2.7–3.5. – (1128).

CIL X 8042, 23 compl.   C·BREXI   litt. cavis
                                           SEN̬ECIO

   C. Brexi Senecio(nis).

In his comment in *CIL*, Dressel admitted being uncertain whether the stamp had been found on a brick or a *dolium*, whether or not it rightly belonged among the stamps of Campania, and whether the fifth letter of the first line was an X, as it appeared to him, or an N, as Marini had reported at *Iscr. dol.* 645. This example on a fragment of brick said to have been discovered at Cumae confirms Dressel's reading of the first line and supports his attribution of the stamp to a Campanian source (see the Introduction, p. 6). C. Brexius Senecio is not otherwise known; the *gentilicium* Brexius to my knowledge occurs only once elsewhere, in a funerary inscription from Dyrrachium (*CIL* III 12308). According to Steinby, this type of stamp (small with impressed letters), though chronologically limited to the first century on urban bricks, is common in all periods outside Rome (*Cronologia*, 19).

4.   Sig. 6.0, 5.4; litt. 1.2 (INTH 1.4), 1.3, 1.5 (A 1.0, S 0.7); teg. 60, 46, 3.5. – (1129).

Novum *CIL* X 8042, 60/1   ▭   YACINTHI
                                             IVLIAE
                                             AVGVSTA^S

     Yacinthi Iuliae Augusta(e) s(ervi).

     = Marini, *Iscr. dol.* 13 compl. This example from the De Criscio collection differs from the variants reported at *CIL* X 8042, 60 in the abbreviation S for *s(ervus)* and confirms the reading of Marini, who knew only a poorly preserved copy of the stamp (see the Introduction, p. 6). A third copy, evidently, discovered at Cumae, was reported by L. Jacono, who recognized at least one other variant not known to Dressel among the 22 examples that he examined in the early 1920's (cf. *NSc* 1926, 222, 230 n. 5, q.v. also for a different interpretation of the text). Bloch has identified the Iulia Augusta who appears in a Roman brick stamp (*CIL* XV 1473) with the daughter of Titus (= *PIR*² F 426; *Indices*, 76), and Marini assigns this stamp also to her (*loc. cit.*); but the slave Hyacinthus probably belonged to the empress Livia, who assumed the title Augusta in A.D. 14 (cf. *PIR*² L 301) and who appears in two other Campanian stamps as the owner of slaves Dama and Abda, or Abdaeus (for the stamp of Dama, see *CIL* X 8042, 41 and Maiuri's drawing at *NSc* 1933, 340; for the stamp of Abda (cf. *CIL* VI 19521) or Abdaeus (cf. *CIL* X 1990), see Marini, *Iscr. dol.* 536; Della Corte, *NSc* 1939, 305; and the photograph in Steinby, *Pompei*, 268, fig. 183 a-b). Members of the imperial family are not mentioned in urban stamps before the time of Caligula, but the numerous copies of the stamps of Hyacinthus discovered throughout the Bay of Naples region—at Cumae, Pozzuoli, Naples, Pompeii, Stabiae, Capri, and Ponza—attest the high productivity of Livia's Campanian brickyards during the early years of Tiberius' reign; and the stamps of Dama and Abda suggest that her *figlinae* may have been active already when Augustus was still alive (for the distribution of Hyacinthus' stamps, see Jacono, *loc. cit.* and D'Arms, *CSS*, 78 and n. 27 with refs.).

5.   Sig. 7.9; litt 1.4; lin. 1, ?; fr. 19+, 14+, 2.5. – (1065).

CIL X 8042, 61 (*)   ◎   IANVARI·A·T·L·

     Ianuari A. T(ati) L(aeti).

     = *AJA* 2 (1898), p. 391 no. 45. This example and numbers **6, 10,** and **11** from the De Criscio collection have previously been published by Walter Dennison, who recorded their provenance as Pozzuoli (cf. *AJA* 2 (1898), pp. 391–392 nos. 45–48). He noted the mark of punctuation after L not reported in *CIL* and followed the completion A. T(ati) L(abeonis ?) tentatively suggested by Dressel at *CIL* X 8042, 99. Two unedited stamps have since revealed the full name of the *patronus* to have been A. Tatius Laetus, but his position in the brick industry, whether *dominus* or *officinator*, remains uncertain (cf. Steinby, *Pompei*, 267). Stamps of his slaves and freedmen are frequent at Pompeii and can be dated by type to the last period of the city. Although no examples that can be positively identified have yet been found outside the Bay of Naples region, the stamps evidently originated farther north, near Sinuessa, where a number of recently discovered fragments of tiles and amphorae attest the presence at one time of an active *officina figuli* (see Steinby, *Pompei*, 267, 271 n. 6) Indeed, a fragmentary stamp discovered at Ostia (*LSO* 1251) may belong to the same A. Tatius Philetus who appears in one of the stamps from Pompeii (*CIL* X 8042, 100; cf. *LSO*, p. 391). The slave Ianuarius is known only from this stamp, which in this copy shows no trace of the *signum* found on its counterpart, *CIL* X 8042, 99 (= **11**), to which it otherwise corresponds in form and formula.

6.   Sig. 6.7; litt. 1.2 (I in IVL 1.0); lin. 1, 1; fr. 43+, 25+, 2.0. – (1066).

Novum *CIL* X 8042, 64/5   O   C IVL·DEYTERI·

        C. Iul(i) Deyteri.

= *AJA* 2 (1898), p. 392 no. 46 (*). According to Dennison, this example on a fragment of a roof tile was "found in 1894 east of the amphitheatre" at Pozzuoli (see the comment at **5**). The I in IVL is irregular, in very low relief and framed by an incised perimeter; the V is only faintly visible (see the photograph). The variant reading Y for V in DEYTERI had already been noted by Jorio and Fusco on stamps which differ in other particulars from this example and which Dressel excluded from *CIL* X (cf. *CIL* X 8042, 64b,d). C. Iulius Deuterius is known only from these stamps, which can be dated on the basis of paleography and form to the second half of the first or the first two decades of the second century.

7.   Sig. 9.6, 3.0; litt. 1.9–2.0; fr. 38+, 30+, 2.9. – (1127).

CIL X 8042, 66   ▭   Q·LEPIDI

        Q. Lepidi.

This example on a fragment of a roof tile from the De Criscio collection is one of many copies of the stamps of Q. Lepidius to have been found in Campania, where the bricks evidently originated (see the Introduction, pp. 6, 8 and n. 29; cf. Steinby, *Pompei*, 267); but several examples have turned up at Rome (cf. *CIL* XV 1237 = **91**), and at least two copies found their way as far north as Florence (cf. *CIL* XI 6682b). According to Steinby (*loc. cit.*), the stamps belong to the early Julio-Claudian period.

8.   Sig. 10.5, 2.3; litt. 1.7 (C prima 0.5, C ultima 0.8, E 0.9); teg. 63, 49.5, 2.8. – (1125).

Novum *CIL* X 8042, 76/7   ▭   Q·MVCCI·ASCLEP

        Q. Mu{c}ci Asclep(iadis).

This example on a roof tile said to have been purchased by De Criscio at Cumae may have come from the same die as a copy published by I. Sgobbo with the text Q MVCI ASCLEP (*NSc* 1926, 234 (Naples); cf. Marini, *Iscr. dol.* 1060). Other variants of *CIL* X 8042, 76 have been found at Majorca (Veny, *AEA* 39 (1966), p. 162 no. 8 (fig. 7) = *CIL* XV *A* 103) and near Rome (*CIL* XV 2401; Camilli, *RendAccLinc* 34 (1979), p. 212 no. 89 (Tav. XII, 89)), but the stamps of Q. Mucius Asclepiades are undoubtedly of Campanian origin (see Dressel's comment at *CIL* XV 2401). The small C in ligature, which appears in the same position on *CIL* XV 2401, here presumably represents a mistake of the carver rather than a genuine variant spelling of the *gentilicium*. Paleography and form suggest a date in the first half of the first century. On the Campanian Mucii, see Castrén, *Ordo*, pp. 192–193 no. 257.

9.   Sig. 11.7, 2.6; litt. 2.2 (R 2.4); fr. 49.5, 49+, 3.8. – (1124).

Novum *CIL* X 8042, 94/5   ▭   SERAPIONIS

        Serapionis.

        This example on a fragment of a roof tile is from the De Criscio collection, for which see the Introduction, p. 6. The name of the slave was Serapio or Serapion: both are well attested in Campania (cf. *CIL* X 2237, 3018, 3238, 3574, etc.). Paleography and form suggest a date in the first half of the first century.

10.   Sig. 9.5, 2.5; litt. 2.2; fr. 19+, 12+, 4.2. – (1068).

Novum *CIL* X 8042, 98/9   ▭   ST·FL M

        St. *vel* St(ati) Fl(avi) M( ) ?

        = *AJA* 2 (1898), p. 392 no. 47 (Pozzuoli; see the comment at **5**). Before the die was impressed a smooth shallow groove was scraped into the wet clay, perhaps by a finger (see the photograph; cf. the Appendix, 1). The horizontal stroke extending off the base of F apparently represents an L in ligature FL (for a similar ligature in a fourth-century stamp, cf. *CIL* XV *S* 602); the short slanting stroke after ST seems to be a mark of punctuation. Perhaps the first two letters stand for the *praenomen* rather than the *gentilicium* Statius, in which case the stamp belongs between *CIL* X 8042, 52 and 53 (for St. vs. Statius, see Schulze, *LE*, 469). The rounded shape of the letters is unusual and suggests an early date, at the end of the Republic or the beginning of the Empire (cf. *LSO* 178 = *CIL* XV 145 (*) = Steinby, *Cronologia*, Tav. I, fig. 1).

11.   Sig. 8.0; litt. 1.5; lin. –, 1; fr. 22+, 19+, 2.8. – (1067).

CIL X 8042, 99 (*)   ◎   SVCCESSI·A·T·L·
                                                  nux pinea

        Successi A. T(ati) L(aeti).

        = *AJA* 2 (1898), p. 392 no. 48 (Pozzuoli; see the comment at **5**). Dennison reported the marks of punctuation not noted in *CIL* but omitted the *signum*, which on this example can be distinguished only with difficulty (see the photograph). The stamp has been found at Pompeii and can be dated by type to the last period of the city. The slave Successus is not otherwise known.

12.   Sig. 9.4; orb. 3.2; litt. 1.2; lin. 1, 2; fr. 19+, 15+, 2.2. – (1138).

CIL XV 3   ☻   CASTRIS PRAETORI·AVG·N̄
                                       protome Martis (vel militis ?) galeati
                                       et loricati ds., pone iaculum hamatum

        Castris praetori(s) Aug(usti) n(ostri).

According to Bloch, the text indicates the destination of the brick rather than its origin (comment at *CIL* XV *S* 2). Evidently Bloch never saw the stamp, which has not been photographed previously, and therefore dated it in his *Indices* only generally, to the second half of the second century (*Indices*, 81). Since on the basis of this example a date in the third rather than the fourth quarter of the century seems fairly certain, the emperor can now be tentatively identified as Marcus Aurelius, or perhaps even Antoninus Pius (see the photograph).

13.   Sig. 9.1; orb. 3.1; litt. 1.2–1.3, 1.3; lin. 1, 2; fr. 12+, 10+, 3.5. – (1142).

CIL XV 20 var. 2   ☉   BRVT M R L MESSAL·ET·PEDONE   a. 115
                                 COS
                        lupus ds. gradiens altero pede anteriore sublato

Brut(iana *sc.* tegula) M. R(utili) L(upi) Messal(la) et Pedone cos.

Cf. *LSO* 36 = *CIL* XV 20 var. 1; *LSO* 34 = *CIL* XV 20a I–II; *LSO* 35 = *CIL* XV 20b. This stamp differs from the versions in *CIL* and *LSO* in the unabbreviated form PEDONE and from *CIL* XV 20a and b also in the lack of ligatures in MESSAL. On M. Rutilius Lupus, perhaps the most important figure in the history of the Roman brick industry, see Bloch, *BL*, 316–320 for the identification of the *dominus* with the prefect of Egypt in the years 113–117 and on the introduction (credited to him) of consular dating on urban stamps in the year 110 (but cf. *CIL* XV *A* 165); further *AJA* 63 (1959), 234–238 on the reasons for the adoption of consular dating; Steinby, *Cronologia*, 27–29, 103 for the attribution and chronology of Rutilius' undated stamps and on his use of the wolf as a rebus; Setälä, *Domini*, 176–180 for the view that Rutilius was both landowner and brick producer and for further prosopographical considerations. Sometime during the first decade of the second century Rutilius Lupus began production at the claylands near the Vatican known as the *figlinae Brutianae*; he continued manufacturing there until his death, probably in the year 123, by which time he had acquired also the *figlinae Naevianae* and *Narn(ienses)* (for the location of the *Brutianae*, see Bloch, *Suppl.*, 8–9).

14.   Sig. 8.8; orb. ?; litt. 0.9–1.0, 1.0, 1.0; lin. –, 2, 2; fr. 14+, 11+, 3.7. – (1143).

CIL XV 25a   ☉   M R L Q AQVILIO NIGRO M REBV[LO]   a. 117
                         APRONIANO COS
                               BRV   v. 3 linea recta

M. R(utili) L(upi) Q. Aquilio Nigro M. Rebu[lo] Aproniano cos. Bru(tiana *sc.* tegula).

Cf. *LSO* 45–47 = *CIL* XV 25b–d. On this example the diagonal stroke of R in APRONIANO is missing because of a break in the die; comparison with the P immediately preceding, which shows an open loop, proves that this is the case and that the letter is not a mistake of the carver (see the photograph). For M. Rutilius Lupus and the *figlinae Brutianae*, see the comment on the preceding stamp.

15.   Sig. ?; orb 4.4; litt. 1.5, 1.4; lin. 1, 2, 2; fr. 15+, 11+, 3.6. – (1141).

CIL XV 29b var. 1 II vel 2 II ?   ☉   [BRV]T[IA]NA
                                              LVPI

[Bru]t[ia]na (*sc.* tegula) Lupi.

Cf. Taglietti, *RendAccLinc* 34 (1979), p. 199 no. 34 (Tav. VII, 34/1) = *CIL* XV 29b compl.;

Steinby, *MemAccLinc* 17 (1974), 78 (Tav. I, fig. 3) = Cozzo, Tav. XXIV, fig. 74 = *CIL* XV 29b var. 1 I; Taglietti, *loc. cit.* (Tav. VII, 34/2) = *GIL* XV 29b var. 2 I (star ? at the beginning of the first line) (= ? Steinby, *loc. cit.* (II) = *LSO* 52). This example differs from the previously noted variants in the number of auxiliary lines and in the alignment of the text. The example published by Taglietti as *CIL* XV 29b compl., which shows an arrow at the beginning of the first line, may be another variant rather than a completion of the version in *CIL*: the fragment preserves only the first part of the text. In any case, the Michigan example differs also from this stamp in the arrangement of the letters (note the position of L). For the dating of *CIL* XV 29a–e and all their variants to the first decade of the second century, see Steinby, *Cronologia*, 27 and n. 7. On M. Rutilius Lupus and the *figlinae Brutianae*, see the comment at **13**.

16.   Sig. ?; orb. ?; litt. 1.3, 1.4; lin. –, ?; fr. 8+, 6+, 2.1. – (1745).

CIL XV *S* 22   [EX·FIGLINIS·CAELIAN[IS]
                                      [C·CASS]I C·F·VET

[Ex figlinis Ca]elian[is C. Cass]i C. f(ili) Vet(eris ?).

     = *CIL* VIII 22632, 13. Bloch describes the form as a "crescent with the horns squared off (like [*CIL* XV] 980)". The stamp, previously known from a single copy discovered in Africa, is now attested also in Rome (see the Introduction, p. 6). The *figlinae Caelianae* are otherwise known only from *CIL* XV 49 of the year 120, but Steinby (*Cronologia*, 30) and Bloch (*ad loc.*) agree with Dressel in dating this stamp by type to seventy years before that time. It is not clear in what order the lines of text were meant to be read, but contemporary stamps of the *figlinae Viccianae* (*CIL* XV 656–673) suggest that the reading of semicircular stamps should generally begin with the outermost line (cf. Steinby in *LSO*, p. 30 and n. 31. Bloch and Dressel transcribe the text with the innermost line first). Helen, who concludes that the *dominus* is not named in first-century stamps, would interpret the text as indicating only that C. Cassius C. f. Vet(us ?) "owned" the brick, but he may also have owned the *figlinae*; he appears in brick stamps only here and is not known from other sources (cf. *Organization*, 52–53).

17.   Sig. 8.7; orb. 4.0; litt. 1.5–1.6 (LI 2.0), 1.4; lin. 1, 2; fr. 15+, 10+, 4.2. – (27004, 1147).

CIL XV 50b   CAE SERVILI GELOTIS EX[·FIG]
                                   ISAVR   v. 2 linea fere recta

Cae(pioniana *sc.* tegula) Servili Gelotis ex [fig(linis)] Isaur(icae).

     Cf. *LSO* 81 = *CIL* XV 50a (*); Van Essen, *Santa Prisca*, p. 312 no. 56/DD–24 = *CIL* XV 50 var. (?). Dressel identified the *domina* as Flavia Seia Isaurica and assigned the stamp to the *figlinae Caelianae*, but the *gentilicium* of the *officinator* ties him rather to Plotia (Servilia) Isaurica, the first known owner of the *figlinae Caepionianae*, to which his stamps therefore rightly belong (see Steinby, *Cronologia*, 31 and nn. 1, 2). Ti. Servilius Gelos worked for his *patrona* during the first years of the second century, when she owned the section of the *figlinae Caepionianae* known as the *de Mulionis* or the *Mulionae* (cf. Steinby, *loc. cit.*; Helen, *Organization*, 79–80. Setälä, *Domini*, 164–165 proposes a connection with the Republican Servilii Caepiones—perhaps the eponymous owners of the *Caepionianae*). For the conclusions about the organization of the Roman brick industry that can be drawn from the statistical rarity of this relationship (*officinator* as freedman of *dominus*), see Helen, *Organization*, 105–109. For the complicated history of the *Caepionianae*, see Steinby, *Cronologia*, 30–33, and, for their possible location, near the town of Horta, see Helen, *Organization*, 80–82 (but cf. the comment at *LSO* 326). The stamp has not been photographed previously.

18.   Sig. ?; orb. ?; litt. 1.5 (L̲I̲ 2.0, T̲I̲ 1.9, I ultima 1.1, G ultima 0.8), 1.0; lin. 1, 2; fr. 17+, 9+, 4.3. – (27005, 1148).

CIL XV 50c   ☾   [⫸·CA̲E̲ SERV]IL̲I̲ GELOT̲I̲S EX FIG
                                     [ISA]V̲R   v. 2 linea fere recta

[Cae(pioniana *sc.* tegula) Serv]ili Gelotis ex fig(linis) [Isa]ur(icae).

See the comment on the preceding stamp.

19.   Sig. ?; orb. ?; litt. 1.5–1.6, 1.0; lin. 1, 1, ?; fr. 19+, 10+, 4.4. – (3065, 1151).

CIL XV 65   ☾   [⫸ T·RAV·PA]MP·EX·F·P·IS ⫷
                                     [CAE]P̲ION
                                          ⫷

[T. Rau(si) Pa]mp(hili) ex f(iglinis) P(lotiae) Is(auricae) Cae]pion(iana *sc.* tegula).

    = Taglietti, *RendAccLinc* 28 (1973), p. 317 no. 39 (= *CIL* XV 65 (*): punctuation after CAEPION) ? Interpretation of the second line is based on comparison with a contemporary stamp of L. Gellius Prudens which corresponds to this example in type (*LSO* 86 = *CIL* XV 56 (*); see the comment in *LSO* and Helen, *Organization*, 34). T. Rausius Pamphilus worked as *officinator* at the section of the *figlinae Caepionianae* called the *Mulionae* during the first decade of the second century, when it belonged to Plotia (Servilia) Isaurica, and for a few years after the *Mulionae* had passed into the hands of Arria Fadilla around the year 112, before transferring in the last years of Trajan to the section owned by C. Curiatius Cosanus; sometime in the early Hadrianic period, before the year 123, Pamphilus left the *Caepionianae* altogether for the claylands known as the *Salarese*, where he finished his career under the *domina* Trebicia Tertulla (on T. Rausius Pamphilus, see Helen, *Organization*, 78, 117, 147; Steinby, *Cronologia*, 30, 31 and n. 9, 84 and n. 4; Setälä, *Domini*, 63. For Plotia Isaurica and the *figlinae Caepionianae*, see the comment at **17**). The stamp has not been photographed previously.

20.   Sig. ?; orb. ?; litt. 1.4, 1.2; lin. 1, 2, ?; fr. 11+, 11+, 3.6 – (1728).

CIL XV *S* 25 corr.   [☾]   [EX] FEGLINI[S ARRIA̲E̲ FAD CA̲E̲]
                                              [C] IVLI L̲[VPION̲I̲S]
                                                 caduceus stans

[Ex] feglini[s Arriae Fad(illae) Cae(pioniana *sc.* tegula) C.] Iuli L[upionis].

    = *LSO* 112, which preserves the *signum* and the ligatures omitted by Bloch in the *Supplement*. This example, though more fragmentary than the copy published in *LSO*, nonetheless preserves in part the auxiliary lines, one of which is noted by Taglietti at *RendAccLinc* 34 (1979), p. 200 no. 38. Bloch's primary source seems to have been the version published by Mancini in *NSc* 1914, 394, where punctuation is printed between words and the only ligature is the A̲E̲ in CA̲E̲; *S* 25 shows the punctuation except between IVLI and LVPION̲I̲S (cf. Steinby's comment at *LSO* 112). According to Bloch, the *officinator* C. Iulius Lupio is to be associated with P. Iulius Lupus, the second husband of Arria Fadilla (= *PIR*[2] I 389); the nature of the relationship, however, remains obscure (*HSCP* 63 (1958), p. 413 n. 34). This stamp, which is among his earliest, dates from the last years of Trajan or the first years of Hadrian, shortly after Plotia Isaurica's share of the *figlinae Caepionianae* had passed into the hands of Arria Fadilla (see Steinby, *Cronologia*, 32 and the comments at **17** and **19**; on Arria Fadilla, mother of the future emperor Antoninus Pius, see Setälä, *Domini*, 62–64; for Iulius Lupio's use of the caduceus as a personal emblem, see Steinby, *Cronologia*, 106).

21.   Sig. ?; orb. 3.9; litt. 1.3, 1.3; lin. 1, 1, 2; fr. 15+, 13+, 4.4. – (1156).

CIL XV 81   ☾   EX PRAED [ARRIAES FADILLAE]
                                            M BA[SSI]

Ex praed(is) [Arriaes Fadillae] M(arci) Ba[ssi].

The *officinator* Statius Marcius Bassus had worked at the section of the *figlinae Caepionianae* known as the *Mulionae* since the first years of the second century, when it belonged to Plotia Isaurica, and before that perhaps at the *figlinae Marcianae*, where his family had begun producing bricks and clay vessels at least as early as the time of Caligula and where they had become the leading family of brickmakers during the first century. This stamp, which has not been photographed previously, can be dated from discoveries *in situ* to shortly before the year 123, nearly a decade after Isaurica's share of the *Caepionianae* had passed into the hands of Arria Fadilla (see Steinby, *Cronologia*, 31, 32 and n. 2; on Arria Fadilla, see Setälä, *Domini*, 62–63; on the Statii Marcii, see Helen, *Organization*, 126–127. For Plotia Isaurica and the *figlinae Caepionianae*, see the comment at **17**).

22.   Sig. 9.2; orb. 3.5; litt. 1.1, 0.9, 1.0; lin. 1, 1, 2; fr. 19+, 12+, 3.6. – (1158).

CIL XV 87   ☾   [EX FIG A]RRIAE FADILLAE
                                            PAETIN·ET APRONIAN   a. 123
                                                     COS

[Ex fig(linis) A]rriae Fadillae Paetin(o) et Apronian(o) cos.

See the comment on the preceding stamp.

23.   Sig. 10.5; orb. ?; litt. 1.0, 1.0; lin. 1, 2, 2; fr. 18+, 15+, 2.7. – (1166, 27012).

CIL XV 134   [☾]   [O DO E]X FAVST AVG FIG RANINIANA[S]   sic
                                      [R]VTILI SACESSI

[O(pus) do(liare) e]x Faust(inae) Aug(ustae) fig(linas) ⌐K¬aniniana[s R]utili Sucessi.

= *LSO* 173. This example, though fragmentary, is better preserved than the copy photographed in *LSO*; the brick is broken and mended. In his comment Dressel rightly mentioned, along with the more obvious scriptural errors—R for K in RANINIANAS and the inverted V in SACESSI—the omission of the word *praedis* before the name of the empress; he then further emended the accusative RANINIANAS to the ablative KANINIANIS (so too Helen, *Organization*, 73). Bloch, on the other hand, considers the stamp to be the earliest in which the origin of the brick is indicated by *ex* with the accusative, a linguistic phenomenon that characterizes a group of stamps from the imperial brickyards of Septimius Severus and Caracalla; he corrects the text by supplying a second *ex* before *fig(linas)* (*Indices*, 10, 101; on the use of *ex* and *de* with the accusative, see the remarks of D. Norberg quoted by Bloch at *Indices*, 11–12). Since the carver is not likely to have written A when he meant I, and since he was evidently prone to errors of omission (the spelling

SVCESSI for SVCCESSI, though not without parallel, probably represents just such an error: cf. the two stamps cited below), Bloch's explanation of the text seems slightly preferable to Helen's (for "Sucessus", see Bloch, *Indices*, 74). The *officinator* Rutilius Successus is otherwise known only from the following brick stamp (*CIL* XV 135), also of the *figlinae Caninianae*, and from a *mortarium* stamp recently discovered in a *villa rustica* at Castel di Guido (*CIL* XV *A* 194). The *Caninianae* came into the hands of Faustina Minor (Augusta 146–175) sometime after the year 155, perhaps at the accession of her husband Marcus Aurelius, and ultimately descended to her son, the emperor Commodus (on the *figlinae*, see Steinby, *Cronologia*, 34–35; cf. Setälä, *Domini*, 91, citing Kuusanmäki).

24.   Sig. 8.6; orb. 2.0; litt. 0.7–0.8, 0.7–0.8; lin. 2, 2, 2; fr. 20+, 14+, 3.3. – (1706).

CIL XV *S* 43 (*)   O DO·EX·PR·IMP·C[OMO]·AVC F·CAN
                                                         RVTILI SVCCESS

O(pus) do(liare) ex pr(aedis) imp(eratoris) C[omo(di)] Auc(usti) f(iglinis) Can(inianis) Rutili Success(i).

     = *CIL* XV 135 corr. (*). The spelling COMO for COMMO(dus), though not without parallel in brick stamps, is rare and perhaps represents a mistake of the carver rather than a genuine linguistic variant (cf. *CIL* XI 6688, 5h; *CIL* XV 1058; 1856 = Marini, *Iscr. dol.* 142, with Marini's comment *ad loc.*); the substitution of C for G (AVC), on the other hand, is not uncommon and presumably reflects the contemporary pronunciation (cf. Bloch, *Indices*, 97–98). This is the last stamp of the *figlinae Caninianae*, where the *officinator* Rutilius Successus had worked since the time of Marcus Aurelius (see the comment on the preceding stamp).

25.   Sig. ± 10.4; orb. ?; litt. 1.1 (O in DOLI 0.9), 1.1; lin. ?, 2, ?; fr. 14+, 13+, 2.5. – (1183).

CIL XV 213   O   OP·DOLI·EX PR D[OMINI N̄·AVG]
                                                         EX FIGLIN F[AVRIANI]S
                                    Roma galeata ss. respiciens s. elata hastam, d. demissa
                                    sceptrum tenet; [pone eam sceptrum simillimum]

Op(us) doli(are) ex pr(aedis) d[omini n(ostri) Aug(usti)] ex figlin(is) F[auriani]s.

     The description of the *signum* is adopted from Steinby's completion and correction of Dressel's interpretation: cf. *MemAccLinc* 17 (1974), 81. This example evidently predates the more complete copy photographed in Steinby, *Cronologia*, Tav. III, fig. 6, where the auxiliary lines are preserved (1, 2, 2), but the text has been split by cracks in the die. Bloch has identified the emperor as Septimius Severus and dated the stamp to the years of his sole rule, 193–198 (*Indices*, 81). As is usual in stamps of the Severan era, the *officinator* is here represented by the *signum*, which serves to identify the section of the *figlinae* for which he was responsible (see the comment at **36**); in this case he remains anonymous. For the *figlinae Faurianae* (= *Favorianae*), which undoubtedly constituted a part of the *figlinae Marcianae* in use near the end of the second century, see Steinby, *Cronologia*, 41–42; Bloch, *BL*, 298; and the comment at **76**.

26.   Sig. ± 9.9; orb. ?; litt. 1.3, 1.1; lin. 1, 2, 2; fr. 14+, 11+, 2.5–3.0. – (1184).

CIL XV 220 (*)   ◐   OP·DOL·EX·PR·AVGG N̠N̠ F̠[IG FA]O̠
                                          RIAN·CALP·VERNA·

                                          ♠

Op(us) dol(iare) ex pr(aedis) Aug(ustorum duorum) n(ostrorum) f[ig(linis) Fa]orian(is)
Calp(etanus) Verna.

= *LSO* 233 (*). This example improves the version in *LSO* only slightly, by the addition of a
triangular interpunct after FAORIAN (cf. also the photograph in Gianfrotta, *Castrum Novum*, p. 126, fig.
271). Dressel and Helen interpret the text differently: cf. *CIL* XV,1, p. 68; Helen, *Organization*, 69. Verna is
among the last of a long line of Calpetani who had worked as *officinatores* at the *figlinae Marcianae* since
the time of Caligula, and at the section of them known as the *Favorianae* since their first appearance in the
last years of Commodus. This stamp dates from the years 198–211, when the *Favorianae* belonged to
Septimius Severus and Caracalla. For the *figlinae*, which evidently derived their name from Verna's
distinguished ancestor, C. Calpetanus Favor, see the comment on the preceding stamp. For the C.
Calpetani, see Helen, *Organization*, 27–31, 128–129 and Steinby, *Cronologia*, 62–63, 65 (cf. *CIL* XV *A*
34–35, 104, 128; Camilli and Taglietti, *RendAccLinc* 34 (1979), p. 191 no. 7, p. 202 no. 46). M. Steinby, "I
bolli laterizi dell'Area Sacra di Largo Argentina" in *L'Area Sacra di Largo Argentina I*, ed. F. Coarelli, I.
Kajanto, U. Nyberg, M. Steinby. Rome 1981, p. 330 nos. 140–144 (Tav. LXXIX, figs. 1–4); hereafter:
Steinby, *Largo Argentina*.

27.   Sig. 9.5; orb. 3.1; litt. 1.3, 1.3; lin. –, ?, ?; fr. 19+, 13+, 3.6. – (1186).

CIL XV 225 compl.   ◑   OP[VS DOL EX PRAE] AVREL CA̠E̠S N
                                          E̠X [FIG FVL SV]C̠ES SER
                                          nux pinea

Op[us dol(iare) ex prae(dis)] Aurel(i) Caes(aris) n(ostri) ex [fig(linis) Ful(vianis) Su]ces(sus)
ser(vus).

CIL XV 225 is completed in the description of the form. For the interpretation of the text, cf. the
preceding stamp. The imperial slave Successus is mentioned by name only in this stamp, which probably
dates from between the years 155 ca. and 161, when Marcus Aurelius was Caesar, but he recurs, to judge
from the *signum*, in *CIL* XV 226, which belongs to the time when Marcus Aurelius had become emperor
(see Steinby, *Cronologia*, 42, 43 and n. 1, 106–107). Helen has recently argued for reestablishing the
identification of Aurelius Caesar with Commodus (Caesar 166–177) and of the Lucilla who is the only other
known owner of the *figlinae Fulvianae* with Commodus' sister, Annia Lucilla; he then tentatively dates this
stamp to the autumn of 169, when the emperor Marcus Aurelius might have transferred to his heir some of
the landed property of Annia Lucilla (*Arctos* 10 (1976), 27–36). But the size of the orbiculus, which extends
into the center of the stamp, would seem to recommend the earlier date (see the photograph; cf.
LSO 239 = *CIL* XV 226).

28.   Sig. 9.5+, 5.0; litt. 1.5, 1.5; lin. 2, 2, 2; fr. 11+, 14+, 4.1. – (1201).

CIL XV 303a (*)   ▭   C·SATRINI!C[ELERIS]
                                          ·EX·FIGLINIS[·MARCIA]

C. Satrini C[eleris] ex figlinis [Marcia(nis)].

The mark of punctuation after SATRINI seems to be a downward pointing arrow (see the photograph). The *officinator* C. Satrinius Celer began his career at the *figlinae Marcianae* producing bricks and, through his slaves, *mortaria* before transferring in the Domitianic age to the *figlinae Castricianae* and *Oceanae*, where he made only bricks (cf. Bloch's comment at *S* 548b; Steinby, *Cronologia*, 63). Copies of *CIL* XV 303b have been found in the Colosseum and can therefore be dated to the time of Vespasian; 303a is identical in form and formula to 303b and presumably belongs to the same period, in the early years of Celer's career. On the C. Satrinii, who worked as *officinatores* from the time of Caligula until the middle of the second century, see Helen, *Organization*, 127 (on Celer, see also pp. 33, 48, 82, 94). For the *figlinae Marcianae*, which were exploited intermittently from the middle of the first century up to the Severan period and then again in the age of Constantine, see Bloch, *BL*, 313–314, 334–337 and Steinby, *Cronologia*, 61–66.

29.   Sig. 3.6+, 3.5; litt. 1.1, 1.2; fr. 12+, 10+, 4.2. – (2407).

CIL XV *S* 83 (*)   ▭   [VALERI·CAT]ONIS
                                             [FIGIL·ST]RAB·

[Q. Valeri Cat]onis [ex figil(inis) St]rab(onianis ?).

     = *LSO* 306 (*) (= *CIL* XV *S* 83 compl.). For the integration of the text, see the comments at *LSO* 305 and 306. Despite some apparent differences in the shapes of the letters, this example seems to have come from the same die as the copy published in *LSO*, where the surface of the brick after STRAB is poorly preserved (see the photographs). The *figlinae Strab(onianae ?)* are known only from this stamp and its counterpart, *CIL* XV 2508, both of which belong, on the basis of their paleography and form, to the middle years of the first century (cf. Steinby, *Cronologia*, 88, q.v. also for the proper location of the stamps between those of the *praedia Staton(iensia)* (*CIL* XV 541) and those of the *praedia Subortana* (*CIL* XV 542ff.)). Q. Valerius Cato recurs in a stamp of the *figlinae Marcianae* approximately datable to the time of Nero, but his position in the brick industry remains uncertain (cf. *LSO* 305 = *CIL* XV *S* 82 compl. (*); Steinby, *Cronologia*, 64).

30.   Sig. 10.3; orb. 3.8; litt. 1.1, 1.1; lin. 1, 2, 2; fr. 23+, 20+, 3.1. – (1206).

CIL XV 354 (*)   ☉   OP·EX PR SABINAE·SABINILLAE DE·
                                            FIG·NEGARIANIS↓
                            protome Vulcani pileati ds.; ante malleus, pone forceps

     Op(us) ex pr(aedis) Sabinae Sabinillae de fig(linis) Negarianis.

     The copy published by Cozzo, Tav. LII, fig. 173 apparently came from the same die as this example, which, unlike *CIL* XV 354, shows punctuation after DE and none after SABINILLAE (see the photographs). Both the *figlinae* and the *domina* are known only from these stamps, which can be approximately dated on the basis of form to the middle of the second century (cf. Dressel *ad loc.*).

31.   Sig. 8.4; orb. 4.0; litt. 1.5, 0.9; lin. 1, 1, –; fr. 20+, 8+, 4.4. – (27031, 1752).

Novum *CIL* XV *S* 92/3 ?   ☉   ⇶PRIISCI GAVIAE AMYLLAE ⇷   sic
                                                    AB NEPT

     Pr{i}isci Gaviae Amyllae ab Nept(uno).

In this stamp, which is really a variant of *CIL* XV 355, the I in PRIISCI seems to appear both in ligature (as in *CIL* XV 355) and separately (as in *CIL* XV *S* 92 = *CIL* XV 355 var. 1: the palmettes omitted by Bloch in the *Supplement* are noted at *LSO* 337 (= *S* 92 corr.); cf. Van Essen, *Santa Prisca*, p. 257, no. 53-I/II = *CIL* XV 355 var. 2). Steinby has suggested, perhaps rightly, that since the Kelsey examples have the same measurements and disposition of the text as the copy of *S* 92 published in *LSO*, they are merely copies of the same stamp impressed from a dirty die. On the other hand, the flared shape of the I in ligature is clearly defined on both examples in the Michigan collection and is apparently identical with that of the I in ligature on the copy of *CIL* XV 355 published at *LSO* 336 (= *CIL* XV 355 corr. (*); see the photographs). Perhaps the Kelsey examples represent an original version of the stamp published at *LSO* 337, where the extra I in ligature has been eliminated by filling in the die (with clay ?). This method of adapting a die is not evidenced elsewhere, however, and is admittedly less probable than the more simple adjustment by additional carving that can be inferred, for example, from comparison of *LSO* 343 (= *CIL* XV 362 (*)) with the following stamp (= Novum *CIL* XV 362/3 (*)) (but cf. Steinby's comment on *CIL* XV 1564 in *MemAccLinc* 17 (1974), 94). The slave Priscus worked as *officinator* for Gavia Amylla (= Hamilla) during the first decade of the second century (cf. Steinby, *Cronologia*, 68), when she owned both the *figlinae Platanianae* and the *figlinae ab Neptuno*, for the location of which, see the comment at *LSO* 336 (for Gavia Amylla, see Setälä, *Domini*, 280).

32.   Sig. 9.2; orb 3.3; litt. 1.0, 1.0; lin. 1, 1, 2; fr. 19+, 19+, 4.4. – (1211).

Novum *CIL* XV 362/3 (*)   ☉   ·C·AQVILI·APRILIS·EX·PAEDI   sic
CAES·BIPIDALE·DOLIA   sic
nux pinea foliis circumdata

C. Aquili Aprilis ex p<r>aedi(s) Caes(aris) bip⸢e⸣dale dolia(re).

The stamp was first published in 1974 by Steinby, who recorded three examples in the Antiquariums of the Forum and Palatine (*MemAccLinc* 17 (1974), 83). A fourth copy, it seems, was subsequently discovered *in situ* in a shop on the Via Nova and was published two years later by E. Monaco, whose transcription of the first line, C AQUILIS APRILIS EX PAEDIS, is evidently mistaken (*RendPontAcc* 48 (1975–1976), p. 310 no. 5). Noticing that the same error in the first line of her examples (PAEDI) had been reported by Dressel, Steinby reasonably supposed that the reading BIPIDALE for BIPEDALE constituted a correction rather than a variation of the version in *CIL*, but the recent discovery at Ostia of several copies with the spelling BIPEDALE has since confirmed the existence of two varieties of *CIL* XV 362 (cf. *LSO* 343). Except for the one letter, the stamps are identical in every respect, including the mark of punctuation before the first line, which can be seen also in the photograph in *LSO*. Steinby now suggests, rightly I think, that the stamp in *LSO* was created when the carver changed the I into an E by cutting three horizontal strokes into the die (see the photographs). C. Aquilius Aprilis, who appears as a slave of Aquilia Sozomena in a stamp from the first decade of the second century, worked as *officinator* at the *figlinae Oceanae* at least until the year 125 (cf. Steinby, *Cronologia*, 69, 70 and n. 1). This stamp, his earliest as a freedman, presumably dates from the first years of Hadrian's principate: copies of the corrected version, *CIL* XV 362, have been found in the Pantheon (cf. Bloch, *BL*, 114; on Aprilis see also Helen, *Organization*, 105).

33.   Sig. 8.4; orb. 3.5; litt. 1.0, 0.9; lin. 1, 2, 2; fr. 14+, 21+, 3.2. – (1213).

CIL XV 368 (*)      EX·FIC·OCEA·MAI·CAES·N̄·OP·DO
                                       Q·PERVSI·PVDE·

Ex fic(linis) Ocea(nis) Mai(oribus) Caes(aris) n(ostri) op(us) do(liare) Q. Perusi Pude(ntis).

= *LSO* 347 (*). The Kelsey example improves the version in *LSO* only slightly, by the addition of marks of punctuation after MAI and at the end of the second line. Datable from discoveries *in situ* to the first years of Antoninus Pius, the stamp is the earliest in which the *figlinae Oceanae Maiores* are mentioned by name and the first of the imperial *officinator* Q. Perusius Pudens, who is otherwise known only from *CIL* XV 369 of the year 148 (cf. Steinby, *Cronologia*, 70 and n. 5).

34.   Sig. ± 11.2; orb. 3.6; litt. 1.5, 1.0; lin. 1, 3, 3; fr. 12+, 14+, 3.9. – (1214).

CIL XV 370   ☾   EX·PR·AVG·N̄·OP·D[OL·FIG·]OCE
                              MA HER·ET·VRBICI

Ex pr(aedis) Aug(usti) n(ostri) op(us) d[ol(iare) fig(linis)] Oce(anis) Ma(ioribus) Her(metiani) et Urbici.

= *LSO* 349 corr. This example evidently came from the same die as the copy published in *LSO*, where the innermost series of auxiliary lines is not reported (cf. also the drawing by Righini, Tav. XI, fig. 2 (Cat. no. 55)). Stamps found recently along the Via Praenestina (cf. Blake-Bishop III, p. 112 n. 66) have suggested to Steinby that Bloch's dating of the stamp to the time of Marcus Aurelius (*Indices*, 81) should perhaps be modified to include the reign of Commodus, as indeed Dressel had originally proposed (cf. *Cronologia*, 70 and n. 6). Hermetianus and Urbicus had worked together at the *figlinae Oceanae Maiores* since the brickyards belonged to Marcus Aurelius and Lucius Verus; neither is encountered in stamps without the other, and theirs is the only partnership of *officinatores* that is known to have lasted for more than a year (see Helen, *Organization*, 115).

35.   Sig ± 9.2; orb. ?; litt. 0.8, 0.8, 0.8; lin. 1, 2, 2, 1; fr. 17+, 12+, 3.7–4.7. – (1223).

CIL XV 400   [☾]   [OP]VS DOLLARE EX·PRA F[AVS]   sic
                                [A]VS N FI PONTIC[VL]   sic
                                [Λ]NIS SFX [PO]   sic
                                 [delphinus ds.]

[Op]us dol⌈i⌉are ex pra(edis) F[aus(tinae) A]u⌈g⌉(ustae) n(ostrae) fi(glinis) Pontic[ula]nis S⌈e⌉x. [Po(mpeius)].

Perhaps the form DOLLARE should be accepted as a genuine linguistic variant, but the frequent errors in the text suggest rather that it represents a mistake of the carver. The *officinator* Sex. Pompeius Heli(us ?) or Heli(odorus ?) first appears in two stamps from one of the brickyards owned by the emperor Marcus Aurelius. Prior to his accession Marcus had transferred to his wife sole ownership of the *figlinae Ponticulanae*, where Pompeius worked sometime during the years 169–176 ca. (see Steinby, *Cronologia*, 72–73). The stamp has not been photographed previously.

36.   Sig. ± 10.0; orb. ?; litt. 1.3, 1.2; lin. 1, 2, 2; fr. 19+, 16+, 4.6. – (1226, 27045).

CIL XV 408c   ↻   OP DOL EX PR M AVRELI AN[TO]
NINI AVG N PORT LIC
aries ds.; superne caduceus
alatus, inferne ad d. crumena

Op(us) dol(iare) ex pr(aedis) M. Aureli An[to]nini Aug(usti) n(ostri) port(u) Lic(ini).

Cf. *LSO* 385–387 = *CIL* XV 408a,b,d; this stamp has not been photographed previously. Dressel observed that the variants of *CIL* XV 408 were frequent in the Baths of Caracalla and rightly dated them to the years of Antoninus' sole rule, 212–217 (*CIL* XV,1, p. 121; cf. Bloch's confirmation at *BL*, 291). According to Steinby, *signa* were used at the brickyards of Domitia P. f. Lucilla and her successors from the time of Antoninus Pius and at virtually all brickyards during the Severan period to indicate the *officina* where the brick was made; they are therefore distinctive and can in many cases be linked with the names of *officinatores* and *figlinae* to reveal the origin of otherwise unattributed stamps (see *Cronologia*, 106–108). Here, for example, since the ram is found only on stamps of the freedman Fulvius Primitivus, and in particular only on those of his from the *figlinae Domitianae Minores*, this stamp can be set apart from the other variants of *CIL* XV 408, which belong on the basis of their *signa* to the *figlinae Publilianae*, and can be assigned to the *Domitianae Minores*, where Primitivus worked after the death in 205 of his *patronus*, the powerful C. Fulvius Plautianus (cf. Steinby, *Cronologia*, 38, 74; for the political ascendancy of Plautianus as revealed in brick stamps, see Setälä, *Domini*, 122–127). In the time of Caracalla both brickyards supplied the *portus Licini*, which had served as a center for the storage and distribution of bricks since the middle years of the second century (see Dressel, *CIL* XV,1, pp. 37, 121; Bloch, *RM* 66 (1959), 200–201; Steinby, *Cronologia*, 73–74. For the connections between the *figlinae Publilianae* and *Domitianae*, see Steinby, *Cronologia*, 78).

37.   Sig. ± 9.8; orb. ?; litt. 1.3, 1.0; lin. 1, 2, 2; fr. 12+, 12+, 3.9. – (1227).

CIL XV 408d   ↻   OP DOL EX PR M [AVRELI ANTO]
NINI AVG N P[ORT LIC]
Victoria ss. respiciens, s. palmae ramum,
d. elata coronam tenet; ad s. ara

Op(us) dol(iare) ex pr(aedis) M. [Aureli Anto]nini Aug(usti) n(ostri) p[ort(u) Lic(ini)].

= *LSO* 387 compl. This example, though more fragmentary than the copy published in *LSO*, nonetheless preserves the auxiliary lines (cf. Righini, p. 92, no. 61 (Tav. XIII, fig. 2) = *CIL* XV 408d/e). See the comment on the preceding stamp.

38.   Sig. 9.9; orb. 2.7; litt. 1.1, 0.9; lin. 1, 2, 2; fr. 18+, 12+, 2.5–3.3. – (1229).

CIL XV 427a (*)   ↻   OP·DOL·EX·FIGL·P[VBL]ILIANIS
PR AEMILIAE·SEVERAE C F
Mercurius ss. respiciens, s. cornucopias et chlamy-
dem, d. crumenam tenet; pro pedibus testudo

Op(us) dol(iare) ex figl(inis) P[ubl]ilianis pr(aedis) Aemiliae Severae c(larissimae) f(eminae).

= *LSO* 397 (*). The first letter is attached to the orbiculus (see the photograph). The mark of punctuation after DOL is reported in *CIL*, the one after EX in *LSO*. Since the figure of Mercury can be distinguished, albeit with difficulty, this example is presumably a worn copy of *CIL* XV 427a rather than a copy of 427b, which Steinby suggests was created by the deliberate cancellation of the *signum* of 427a (*Cronologia*, 77 n. 6). For the identification of the *domina* Aemilia Severa and the dating of her stamps to the time of Septimius Severus, see Steinby, *Cronologia*, 75 and n. 13, 76, 77 and Setälä, *Domini*, 50–52. The *signum* recurs in a stamp of the emperor Caracalla (*CIL* XV 408b), who acquired the *Publilianae*, perhaps at the death of Septimius Severus, from Severa's adolescent son, Flaccus Aelianus, on whom see Setälä, *Domini*, 111–112 (see also the comment at **36**).

39. Sig. ?; orb. ?; litt. 1.0, 1.0; lin. 1, 2, ?; fr. 16+, 9+, 3.0–3.8. – (1230).

CIL XV 433 (*) [◯] [O]P̣VS·DOLI[ARE EX·PRAEDIS]
[AE]MILỊ[AES·SEVERES]
[piscis ss.]

[O]pus doli[are ex praedis Ae]mili[aes Severes].

Identification of this example with *CIL* XV 433 is based on comparison with the drawing by Righini, Tav. XIV, fig. 1 (Cat. no. 65), where the mark of punctuation after OPVS is omitted, and with the photograph in Cozzo, Tav. LVIII, fig. 196, where the surface of the brick is poorly preserved. For Aemilia Severa and the attribution of her stamps to the *figlinae Publilianae* of the Severan period, see the comment on the preceding stamp. The anonymous *officinator* represented here by the *signum piscis* is not otherwise known, but the emblem recurs on a fragmentary stamp tentatively assigned by Steinby to the *figlinae Publilianae* of Caracalla (*CIL* XV 1800; see Steinby, *Cronologia*, 77 n. 11 and the comment at **36**). The same *signum* was used at the end of the second century at the *figlinae Propetianae* when they belonged to M. Cassius Hortensius Paulinus (cf. *CIL* XV *S* 105), and by an imperial *officinator* S( ) St( ) ? sometime in the years 198–211, perhaps at the same brickyards, under the emperors Septimius Severus and Caracalla (cf. *CIL* XV 770; Bloch, *BL*, 302. For the *Propetianae*, which are not otherwise known to have been an imperial property, see Steinby, *Cronologia*, 75; for Hortensius Paulinus, see Setälä, *Domini*, 129–131).

40. Sig. ± 8.8; orb. ± 3.4; litt. 1.3–1.4; lin. 1, 2; fr. 17+, 12+, 3.7. – (27046, 1232).

CIL XV 451 ◯ L ATTI CHẠ[RITONIS Q]

L. Atti Cha[ritonis Q(uintanense *sc.* opus)].

= *LSO* 412. Only one of the Kelsey examples (27046) shows traces of the second line of text reported by Steinby at *LSO* 412: a horizontal stroke, perhaps the base of an E or an L, can be seen above the AT of ATTI; two vertical strokes (of an N ?) are visible above the H in CHARITONIS. The horizontal stroke of L is fragmentary because of a break in the die (see the photograph). The *officinator* L. Attius Charito is known only from this stamp, which Bloch has dated from discoveries *in situ* to the first years of Hadrian (cf. *BL*, 150, 197; *Ostia* I, 223 (Is. X, 3), 224 (Is. X, 2)). On the distinction between the *figlinae Quintianae* and the *praedia Quintanensia*, to which this stamp belongs, and for the location of both, see Bloch, *BL*, 204–210 and *Suppl.*, 30. For the history of the ownership of the *praedia Quintanensia* and discussion of the problems involved in attributing this stamp and others to the various *domini*—M. Annius Verus, his wife Rupilia Faustina, and the imperial freedman Agathyrsus—see Steinby, *Cronologia*, 78–80.

41.   Sig. 8.8; litt. 1.6; fr. 19+, 6+, 4.0. – (27126).

CIL XV *S* 115   M A V Q   litt. cavis

   M. A(nni) V(eri) Q(uintanense *sc.* opus).

   The die was impressed twice (see the photograph). Although this example differs from the fragment published at *LSO* 421, the latter cannot yet be positively identified with *CIL* XV *S* 116, which evidently came from various dies: see Van Essen, *Santa Prisca*, p. 286 no. 55/T-23 and the copy photographed at Pl. LXXI, 6 (= p. 314 no. 56/EE-3), which does not match the example from Ostia. This stamp, datable by type to the last years of Trajan or the first years of Hadrian, is among the earliest of the *dominus* M. Annius Verus, whose properties included the *figlinae Germ(anicianae)* and a section of the *Salarese* as well as a part of the *praedia Quintanensia*, for which see the comment on the preceding stamp (for the name of the *Germ(anicianae)*, see the comment at *LSO* 250; for the *Salarese*, see the comment at **44**). Three times consul, Annius Verus was undeniably the most prominent senator of his generation; and as *praefectus urbi* in the years 121–125, he may well have influenced the emperor's decision to institute universally the consular-dating of brick stamps in the year 123 (see Steinby, *Cronologia*, 80, and, further on Annius Verus, Setälä, *Domini*, 56–58; for M. Rutilius Lupus and the introduction of consular dating on urban stamps in the year 110, see the comment at **13**).

42.   Sig. ± 10.6; orb. ?; litt. 1.2, 1.2–1.3, (S 1.0, O 0.7, Ɔ 1.1); lin. 1, 2, 2; fr. 23+, 19+, 3.2–4.5. – (1236).

CIL XV 468a   ☾   [EX PR AGA]THYRSI AVG LIB
                                    [GLA]B ET HOMVL   a. 152
                                    COƆ   linea recta

   [Ex pr(aedis) Aga]thyrsi Aug(usti *vel* –ae) lib(erti) [Gla]b(rione) et Homul(lo) cos.

   For a drawing of the entire stamp, which has not been photographed previously, see Righini, Tav. XIV, fig. 2 (Cat. no. 69). According to Bloch, M. Annius Verus sold the *praedia Quintanensia* to Agathyrsus, the freedman of Trajan or Plotina, in the year 134; but Steinby's theory that the *praedia* had been divided even before the year 123 more closely conforms to the evidence, which suggests that Agathyrsus was active already during the first years of Hadrian's reign (cf. Bloch, *BL*, 209–210; Steinby, *Cronologia*, 80. For Annius Verus and the *praedia Quintanensia*, see the comment on the preceding stamp). Indeed, there is some reason to believe that the senator and the imperial freedman collaborated in managing their brickyards: the workshops located in Agathyrsus' section of the *Quintanensia* seem to have produced mainly the large bricks—*tegulae, bipedales,* and *sesquipedales*—whereas those on the property of Annius Verus specialized in *bessales* (see Steinby, *loc. cit.* and, on Agathyrsus, Setälä, *Domini*, 52–54). The division of the *Quintanensia* may have occured at the death of Verus' wife, Rupilia Faustina, whose stamps suggest that the installations on her property manufactured both types of bricks, but since neither Verus' nor Agathyrsus' *figlinae* produced the large or the small bricks exclusively, and since the precise date of Faustina's death is unknown, the early history of the *Quintanensia* necessarily remains obscure (for Rupilia Faustina, see Setälä, *Domini*, 174–175).

43.   Sig. 9.6; orb. 4.7; litt. 1.5; lin. 1, 2; fr. 18+, 15+, 3.7. – (1237).

CIL XV 473   ☾   ⟫⟫ VLPI·ANICETI·EX·FIG·ÇAES·N̄ ⟪⟪

Ulpi Aniceti ex fig(linis) Caes(aris) n(ostri).

= *LSO* 433, which is fragmentary. It is just possible that the *officinator* M. Ulpius Anicetus who appears in this stamp and in *CIL* XV 472, both from the *figlinae Rhodinianae* of the Trajanic or early Hadrianic period, is the same as the Ulpius Anicetianus who was active at the *figlinae Terentianae* at least as late as the year 155, as Dressel first tentatively proposed and indeed as Helen and Setälä have accepted; but Steinby and Bloch are probably right in considering the two to be different persons (see Setälä, *Domini*, 146 and Steinby's comment at *LSO* 433, with further references).

44.   Sig. 7.6+, 3.0; litt. 1.3, 1.3; fr. 20+, 11+, 3.7. – (1239).

CIL XV 486a   [APR ET P]AET COS EX F   a. 123
                                 [COR MA]LL SAL P P B   litt. cavis

[Apr(oniano) et P]aet(ino) cos. ex f(iglinis) [Cor(neliae) Ma]ll(iolae) Sal(arese *sc.* opus) P. P( ) B( ).

= *LSO* 445. This example preserves a portion of the stamp not shown in the photograph in *LSO*. Bloch has identified the *domina* Cor(nelia) Malliola (= Manliola) with the dedicand of a sepulchral inscription erected at Tibur to the consular *dominus* (Cn. Pinarius) Cornelius Severus (*CIL* XIV 3604), but the nature of the relationship between the two remains uncertain (*BL*, 180 and n. 51). Since the stamps of Manliola were consular-dated in the year 123, whereas those of Cornelius Severus belong to the early Trajanic period, Setälä's suggestion that Severus and Manliola were related as father and daughter is attractive (*Domini*, 102–103). If the letters P P B are indeed the initials of an otherwise unidentified *officinator*, the fact that he worked at the brickyards of at least four different *domini* in the year 123 would support Helen's view that it was the *officinator* rather than the *dominus* who was the entrepreneur in the brick industry (see Helen, *Organization*, 116–118; cf. Steinby, *Cronologia*, 83 and n. 4). For the complicated history of the claylands along the Via Salaria that yielded the products known as *opus Salarese*, see Steinby, *Cronologia*, 83–86.

45.   Sig. 11.9+, 3.0; litt. 1.4, 1.4; fr. 21+, 14+, 3.6. – (1242).

CIL XV 507b var. vel (*)   MARCELL ET CELS II COS   a. 129
                                 [EX PR· VLP·] VḶPIANỊ SAL   litt. cavis

Marcell(o) et Cels(o) II cos. [ex pr(aedis) Ulp(i)] Ulpiani (?) Sal(arese *sc.* opus).

Cf. Dressel at *CIL* XV 507 no. 13: "In hoc exemplo dubitari potest utrum VLPIAN· an VLPIANI scriptum sit." The latter interpretation, I think, is preferable: the mark has the shape of an I, the bottom part of which is missing because of a break in the die (see the photograph). On the *dominus* M. Ulpius Ulpianus, whose stamps all were consular-dated in the middle years of Hadrian's reign (aa. 123–133), see Bloch, *AJA* 63 (1959), p. 237 n. 46 and Setälä, *Domini*, 205. For *opus Salarese*, see the comment on the preceding stamp.

46.   Sig. 12.8+; litt. 2.0; fr. 14+, 11+, 3.5. – (1245).

CIL XV 519   EX PR·T·F AMP̣[·B S]   litt. cavis

Ex pr(aedis) T. F(lavi) Amp(liati) [b(essalis) S(alaresis)].

The lower half of P in PR is missing because of a break in the die, which was impressed twice. The marks of punctuation are only faintly visible; cf. Dressel *ad loc.*: "interpunctionis signa non in omnibus exemplis extant" (see the photograph). For the interpretation *b(essalis) S(alaresis)*, see Dressel's comment in *CIL* and Steinby's (indirect) confirmation at *Cronologia*, 19. Steinby has identified the T. F(lavius) Ampliatus who appears in this stamp as an owner of brickyards producing *opus Salarese* during the Trajanic or early Hadrianic period with the T. F( ) A( ) known from a Trajanic stamp of the *figlinae Sulpicianae* (*Cronologia*, 84, 90; cf. p. 91, where it is noted that at least two other *domini* of the period produced both *opus Sulpicianum* and *opus Salarese*, for which see the comment at **44**).

47.   Sig. 13.9, 3.3 (marg. vid.); litt. 2.1; fr. 17+, 10+, 4.0. – (1720).

CIL XV 525c III vel IV   SAL EX PR /RẸ   litt. cavis

Sal(arese *sc.* opus) ex pr(aedis) [[T]]re(biciae).

Dressel noted that *CIL* XV 525c had been impressed from various dies, of which at least two varieties have now been identified among the stamps from Ostia: cf. *LSO* 462 = *CIL* XV 525c I–II; *LSO* 464 = *CIL* XV 529 or 525c III. This example differs from the previously noted variants primarily in the length of the stamp and in the height of the letters. Since the T is missing and the R in PR is fragmentary because of defects in the die, the final letter is presumably a broken E rather than an I (see the photograph; but, on the substitution of I for E, see Bloch, *Indices*, 98). The variants of *CIL* XV 525, datable from discoveries *in situ* to the first years of Hadrian, are among the earliest stamps of Trebicia Tertulla, who in the year 123 was one of at least three landowners whose brickyards produced *opus Salarese* (see Bloch's comment at *S* 142; Steinby, *Cronologia*, 84 and n. 3; Setälä, *Domini*, 195; and the comment at **44**).

48.   Sig. 10.3; orb. 3.4; litt. 1.1, 1.2 (N̲I̲ 1.4); lin. 1, 2, 2; fr. 18+, 15+, 2.5. – (94940).

CIL XV 533a   ☺   EX·PRA̲E̲ AVG·N̄ DE FIGL·SEF·OP
                                   P·MELLVTI·EVVEN̲I̲
                             vas utrimque ansatum ds. iacens

Ex prae(dis) Aug(usti) n(ostri) de figl(inis) Sef( ) op(us) P. Melluti Euveni.

Cf. Righini, p. 110 no. 77 (Tav. XVI, fig. 1) = *CIL* XV 533b. Since the *officinator* P. Mellutius Euvenus and, more important, the *figlinae Sef*( ) are known only from the two variants of *CIL* XV 533, Dressel's suggestion that the carver mistakenly wrote F for P in SEF and that the stamps belong to the *figlinae Septimianae* must be considered entirely plausible (*CIL* XV,1, p. 153; for the *Septimianae*, an imperial property at least since the Hadrianic period, see Steinby, *Cronologia*, 87; for the abbreviation *Sep.*, cf. *CIL* XV 537b). The shape of the letters and the size of the orbiculus led Dressel to identify the emperor as Marcus Aurelius or Commodus, but the stamp is perhaps datable to as early as the time of Antoninus Pius: a copy has been found at Ostia in a building approximately dated to the middle of the second century (cf. Steinby, *Cronologia*, 86 and n. 6).

49.   Sig. 9.1; orb. 3.9; litt. 1.1; lin. 1, 1; fr. 20+, 20+, 4.6–5.0. – (1248).

CIL XV 556      EX FIGLINIS·DOM DOM

Ex figlinis Dom(itiae) Dom(itiani).

A fingerprint is impressed in the orbiculus (see the photograph; cf. Bloch, *RM* 66 (1959), 196 (Taf. 59, 1)). Domitia Domitiani is to be identified with the widow of Domitian, Domitia Longina (= *PIR*[2] D 181; see Setälä, *Domini*, 109–111). She first appears in brick stamps as an owner of *figlinae Sulpicianae* in the early Hadrianic age, to which Steinby assigns this stamp on the basis of its form, and she was active at least as late as the year 127, when she must have been nearly ninety years old (cf. *Cronologia*, 90). During the Trajanic and early Hadrianic periods at least nine persons owned claylands known as *Sulpicianae* in the region where a century later the emperor Caracalla built his baths (on the *figlinae*, see Steinby, *Cronologia*, 89–91, and for their location, *RE* Supplb. XV, 1509).

50.   Sig. ± 9.2; orb ?; litt. 1.1–1.2, 1.0; lin. 1, 2, 2; fr. 14+, 18+, 2.7. – (27057, 1250).

CIL XV 561      VILLI ALEXANDR[I]
                                              SVLPICES

Villi Alexandr[i] Sulpices(e *sc.* opus).

A. Villius Alexander is otherwise known only from *CIL* XV 560 of the year 123, where he appears as an owner of *figlinae Sulpicianae*; this stamp, which has not been photographed previously, can be dated on the basis of form to shortly before that time (cf. Steinby, *Cronologia*, 91; Setälä, *Domini*, 202. For the *Sulpicianae*, see the comment on the preceding stamp).

51.   Sig. 13.5, 3.5; litt. 1.6, 1.6–1.7; fr. 19+, 10+, 3.8. – (29525).

CIL XV 563c compl.   PAETIN ET APRONIA   a. 123
                               M VINIC PANTG SVLP   sic   litt. cavis

Paetin(o) et Apronia(no *sc.* cos.) M. Vinic(i) Pant<a>g(ati) Sulp(icianum *sc.* opus).

Cf. Righini, p. 118 no. 85 (Tav. XVI, fig. 2) = *CIL* XV 563f; Steinby, *MemAccLinc* 17 (1974), 84 = *CIL* XV 563r corr. This example from the Van Deman collection preserves in its entirety a stamp that Dressel knew only in fragmentary form. The die was impressed twice: to the left of lines 1 and 2 are an A and an M respectively that bear the same spatial relationship to each other as the A in PAETIN and the M in the second line. Since many of the letters are fragmentary because of breaks in the die, the P in SVLP, which resembles an F, presumably represents a similar defect rather than a mistake of the carver; cf. Dressel *ad loc.*, no. 10: "postrema littera v. 2 F potius quam P videtur esse, casu opinor". Bloch has argued convincingly that the eighteen variants of *CIL* XV 563 were the personal stamps of slaves in the service of M. Vinicius Pantagatus in the year 123 (*BL*, 322); but whether Pantagatus himself, possibly a freedman of M. Vinicius Crescens, was *dominus* or *officinator* at the *figlinae Sulpicianae* remains uncertain (cf. Steinby, *Cronologia* 91 and nn. 4 and 5; for Crescens, see *LSO* 1096 = *CIL* XV 1527 corr.; *CIL* XV 1528). Two other M. Vinicii, Herculan(us) and Fortunat(us), appear in stamps of the year 123, the latter certainly as *officinator*, and a third, Salvianus, was active at the *Sulpicianae* during the late Trajanic and early Hadrianic periods, but the nature of their connections with Pantagatus cannot yet be determined (for

Fortunat(us), see the comment at **101** (= *CIL* XV 1414); Herculan(us): *CIL* XV 1529; Salvianus: cf. Steinby, *Cronologia*, 91 and n. 6. For the *figlinae Sulpicianae*, see the comment at **49**).

52.   Sig. 5.2.+, 3.7; litt. 1.8, 1.7; fr. 10+, 14+, 3.8. – (1253).

CIL XV 563i   [PAETIN ET AP]RONIA   a. 123
                                  [M VINIC PAN]TAG SVL   litt. cavis

[Paetin(o) et Ap]ronia(no *sc.* cos.) [M. Vinic(i) Pan]tag(ati) Sul(picianum *sc.* opus).

The stamp is poorly preserved in this example but has not been photographed previously. See the comment on the preceding stamp and, for the *figlinae Sulpicianae*, the comment at **49**.

53.   Sig. 9.3+, 6.5; litt. 3.1, (J 2.6, O 2.2); fr. 12+, 9+, 3.4. – (1254).

CIL XV 565e   ⅃VƧ N]Vꟼ NIV   litt. cavis
                                       O⅃

Vin(ici) Pa[n(tagati) Sul(picianum *sc.* opus)] Jo( ).

Cf. Camilli, *RendAccLinc* 28 (1973), p. 305 no. 16 (Tav. III, fig. 2) = Steinby, *MemAccLinc* 17 (1974), 84–85 = *CIL* XV 565m compl. For the interpretation of the letters in the second line as representing the name of the slave who made the brick—here, perhaps, Jo(vinus) (cf. Kajanto, *LC*, 212)—and on the dating of the entire series *CIL* XV 565a-n to about the year 120, see Bloch, *BL*, 322. On M. Vinicius Pantagatus, see the comment at **51**; for the *figlinae Sulpicianae*, see the comment at **49**. The stamp has not been photographed previously.

54.   Sig. 8.7+, 4.2; litt. 1.3, 1.3, 1.2; fr. 15+, 14+, 3.2. – (1257).

CIL XV 575   EX FIGLI[NIS]   litt. cavis
                                  CAECIL QV[INTAE]
                                  SVLP[ICIANI]

Ex figli[nis] Caecil(iae) Qu[intae] Sulp[iciani(s)].

Whether there was a mark of punctuation on this example between CAECIL and QV[INTAE], as in Marini, *Iscr. dol.* 661, or not, as in *CIL*, cannot be determined, because the surface of the brick is poorly preserved; cf. Dressel *ad loc.*: "Hoc sigillum diversis signaculis impressum est." The *domina* Caecilia Quinta is known only from three stamps of the *figlinae Sulpicianae*, all of which can be dated by type to the early Hadrianic period (cf. Steinby, *Cronologia*, 90 and n. 3; Setälä, *Domini*, 82. For the *figlinae*, see the comment at **49**).

55.   Sig. 12.0+; litt. 3.3; fr. 22+, 19+, 3.7. – (93923, 1259, 27065).

CIL XV 585b (*)   C·CVL DIA SVL   litt. cavis

C. Cul( ) Dia(dumeni ?) Sul(picianum *sc.* opus).

Steinby has identified at least ten different stamps pertaining to *CIL* XV 585b, several of which show the punctuation noted here after the *praenomen* (cf. *MemAccLinc* 17 (1974), 85; *Battistero*, p. 124 no. 60 = 585b var. ?: SVL̲P ?). The three examples in the Michigan collection evidently came from the same die. C. Cul(   ) Dia(dumenus ?) is known only from the variants of *CIL* XV 585, of which b and c can be dated from discoveries *in situ* to the second decade of the second century (cf. Bloch, *BL*, 211–217; Steinby, *Cronologia*, 90 and n. 11). For the *figlinae Sulpicianae*, see the comment at **49**.

56.   Sig. 8.7+, 6.3; litt. 2.1, 2.3; fr. 18+, 13+, 4.2. – (3062).

CIL XV 591a   [SV]L̲P   litt. cavis
                                                [P]OP·EVN

       [Su]lp(icianum *sc.* opus) [P]op(   ) Eun(   ).

       On this example the lowest horizontal stroke of E is missing because of a break in the die (see the photograph). The unidentified Pop(   ) Eun(us ?) perhaps recurs in *CIL* XV 1991, a fragmentary stamp which, as the text is printed in *CIL*, would reveal his *praenomen* to have been Publius; but *CIL* XV 1991 is probably just a poorly preserved copy of 591b. This stamp, which has not been photographed previously, has been found in the Baths of Trajan and can therefore be dated to the years of their construction, 104–109, or shortly before (cf. Bloch, *BL*, 41, 45; Steinby, *Cronologia*, 91). For the *figlinae Sulpicianae*, see the comment at **49**.

57.   Sig. 7.2+, 2.5 (marg. vid.); litt. 2.5; fr. 14+, 11+, 3.2–3.4. – (1261).

CIL XV 593 (*)   C·VILLI CṚ[ES SV̲L]   litt. cavis

       C. Villi Cr[es(centis) Sul(picianum *sc.* opus)].

       For the punctuation, cf. *CIL* XV 593 no. 12. C. Villius Cres(cens) is known only from this stamp, which Steinby has dated from discoveries *in situ* to shortly before the year 123 (*Cronologia*, 91; cf. Bloch, *BL*, 59). For the *figlinae Sulpicianae*, see the comment at **49**.

58.   Sig. 7.6+; litt. 2.3; fr. 13+, 11+, 3.4. – (1265).

CIL XV 598a corr.   [R]AB SVḶ[P]   litt. cavis

       [R]ab(   ) Sul[p(icianum *sc.* opus)].

       This example, in which the letters are impressed and the upper loop of B is visible, probably constitutes a correction rather than a variant of the stamp in *CIL*, which Dressel did not see. The diagonal stroke over V seems not to be part of a letter (see the photograph). According to Steinby, all stamps of the *figlinae Sulpicianae* that consist of only a cognomen and the abbreviation *Sul.* or *Sulp.* belong to the beginning of the second century (*Cronologia*, 89). For the *figlinae*, see the comment at **49**.

59.   Sig. 11.9+, litt. 2.2 (I 1.9); fr. 14+, 10+, 3.6. – (1266).

CIL XV 599c   RVFI SV[L]   litt. cavis

Rufi Su[1(picianum *sc.* opus)].

Cf. *LSO* 495 = *CIL* XV 599b. Identification of this example with *CIL* XV 599c rather than 599a is based on the size of the letters, which Dressel described as "larger than usual". On the form and date of the stamp, which has not been photographed previously, see the comment at **58** (cf. Bloch, *BL*, 48). For the *figlinae Sulpicianae*, see the comment at **49**.

60.   Sig. 8.8; orb. 3.8; litt. 1.5, 1.0; lin. 1, 2, –; fr. 21+, 21+, 3.9. – (27067, 1268, 1739).

CIL XV 609a   ☻   C·CALPVRNI·SPATALI
                                            TEMPESINA

C. Calpurni Spatali Tempesina (*sc.* tegula).

Unlike *CIL* XV 609a, which has not been photographed previously, *LSO* 519 (= *CIL* XV 609b (*)) and 520 (= *CIL* XV 609c (*)) show the center of the stamp in relief. According to Steinby, C. Calpurnius Spatalus should be considered a freedman-*officinator* of Calpurnia Secunda, *domina* of the *figlinae Tempesinae* during the Trajanic era, because his stamps are contemporary with hers (*Cronologia*, 92 and n. 5; cf. Setälä, *Domini*, 85). But the fact that a stamp of Calpurnia (*CIL* XV 611; cf. *LSO* 522) is identical in form and formula with this stamp of Spatalus suggests that the two held the same position in the brick industry, that, as Bloch has proposed, Spatalus was a relative of hers and *dominus* in his own right (*BL*, 56. Helen, writing before the publication of Steinby's monograph, does not include Spatalus in his list of *officinatores* who were freedmen of *domini*: cf. *Organization*, 107). Even if the text of *CIL* XV 609a were meant to be read with the innermost line first, as the other two variants of 609 suggest, and even if we see the stamp as indicating only that Spatalus "owned" the brick, even then the text could not be said to support the view that he was merely an *officinator* at the *Tempesinae* (cf. Helen, *Organization*, 90–91). Spatalus is known only from these three stamps, which Bloch and Steinby agree in dating to the first decade of the second century (*loc. cit.*). For the history of the ownership of the *figlinae Tempesinae* after that time, see Steinby, *MemAccLinc* 17 (1974), 98–100 and *Cronologia*, 92–93.

61.   Sig. 9.7; orb. ?; litt. 1.2, 0.9; lin. 1, 2, ?; fr. 18+, 15+, 3.7. – (1270).

CIL XV 630b (*)   ☻   EX F·TERNT·DOM·LVC·POR[T]   sic
                                        LIC·OP DOL·STAT·PRIM
                                        discus (luna plena ? sol ?)

Ex f(iglinis) Ter<e>nt(ianis) Dom(itiae) Luc(illae) por[t(u)] Lic(ini) op(us) dol(iare) Stat(iae) Prim(ulae).

Cf. *LSO* 533 = *CIL* XV 630a (*). The E in TERNT is in very high relief, ca. 0.3 cm. Dressel rightly interpreted the *signum* as a celestial body from comparison with other stamps of Statia Primula (= Primilla) that show a crescent moon and stars at the center (*CIL* XV 139, 630a). The *figlinae Terentianae* are unattested before the early years of Antoninus Pius, when they appear among the properties of Domitia P. f. Lucilla, mother of the future emperor Marcus Aurelius (= *PIR*[2] D 183). From this period date the stamps of the *officinatrix* Statia Primula, who produced bricks also at Domitia's *figlinae Caninianae* (see Steinby,

Cronologia, 93–94; cf. *CIL* XV 139–140). On the *gens Domitia*, the most prominent family in the history of the Roman brick industry, whose 250 stamps dating from the time of Caligula up to the death of the emperor Commodus provide an unprecedented view of family continuity in business in the ancient world, see, most recently, Helen, *Organization*, 100–102; Steinby, *Cronologia*, 47–58; and Setälä, *Domini*, 107–109. For the stamps of the younger Domitia Lucilla, see also A. C. G. Smith, *PBSR* 33 (1978), 82–87. For the *por(tus) Lic(ini)*, see the comment at **36**.

62.   Sig. 9.0; orb. 5.9; litt. 1.0–1.4, 0.7–1.2; lin. –, 1, –; fr. 18+, 10+, 2.2. – (1271).

CIL XV 633a   ☾   ⟫⟩TONNEIANA·ZOSIMI ⟨⟨⟪   vers. falc.
                                            [←] L·IVLI·RVFI →

        Tonneiana (*sc.* tegula) Zosimi L. Iuli Rufi.

        Cf. *LSO* 537–538 = *CIL* XV 633b-c. This example may be identical with the copy published at *LSO* 536, which lacks a photograph, or it may have come from one of the various dies reported in *CIL*. Despite the reservations of Dressel (comment at 633), it is virtually certain that L. Iulius Rufus, consul in 67, was *dominus* during the Vespasianic era, and that his slave Zosimus worked for him as *officinator* (see Bloch, *Indices*, 34; Setälä, *Domini*, 39–40; Steinby, *Cronologia*, 97 and n. 3). Interpretation of the term "Tonneiana", on the other hand, has proved difficult since it occurs in stamps in connection with the names of no fewer than four other *figlinae*. It is generally agreed that the *figlinae Tonneianae* derived their name from a member (or members) of the Tonnei family who were active at the *figlinae Viccianae* near the middle of the first century. Bloch suggests that the Tonnei may have invented "some sort of improvement in the production of bricks" and that other manufacturers subsequently adopted the name along with the process (comment at *S* 202; see also the remarks at *S* 109 and *S* 196). Against this theory Steinby proposes that the *Tonneianae* constituted a unity of which the other *figlinae* formed various subsections (*Cronologia*, 97–98, 105). She points out that at least until studies on the location of the *figlinae* near Rome prove or disprove her hypothesis, the meaning of the term "Tonneiana" will remain obscure (*Cronologia*, 98).

63.   Sig. 9.6; orb. 3.6; litt. 1.1, 1.0; lin. 1, 2, 2; fr. 9+, 17+, 3.9–4.2. – (1273).

CIL XV 648b   ☾   TEG DO[L DE F]IG IVLIAE PROCVLA
                                            PAETINO·COS   a. 123
                                    modius ex quo pendent spicae tres

        Teg(ula) do[l(iaris) de f]ig(linis) Iuliae Procula(e) Paetino cos.

        = *LSO* 550; cf. *LSO* 549 = *CIL* XV 648a compl. (*). On this example, which preserves a portion of the stamp not shown in the photographs in *LSO* or Cozzo, Tav. LI, fig. 167, the D of DOL is almost entirely missing because of a break in the die (see the photograph). Iulia Procula owned a section of the *figlinae Tonneianae* from the end of the first decade of the second century until at least the year 123, after which time the *figlinae* are unattested before the 140's (see Steinby, *Cronologia*, 99–100). She has been variously identified as the sister or daughter of C. Iulius Proculus, consul in 109, or as the grandchild of L. Iulius Rufus; and Dressel supposed, because of a *signum* and cognomen in common, that she was identical with the Flavia Procula whose stamps were consular-dated in the years 123 and 140 (see the comment at *CIL* XV 1157 and Setälä, *Domini*, 139–140 with further references; for Iulius Rufus and the *figlinae Tonneianae*, see the comment on the preceding stamp. See also Steinby, *Cronologia*, 91 for Procula's connections with the *figlinae Sulpicianae*). For chronological reasons Dressel's identification is improbable, but of course the discrepancy in *gentilicia* does not in itself preclude the identity of the two women: compare

the interesting case of the half sister of Antoninus Pius, Iulia Fadilla, who appears in stamps also as Iulia
Lupula and even as Arria Lupula (cf. Bloch, *HSCP* 63 (1958), 408–411; Helen, *Organization*, 79 and
Righini, p. 36; Steinby, *Cronologia*, 32–33). In any case, Steinby's suggestion that Iulia and Flavia were
related as mother and daughter is more attractive than Dressel's identification of the two (cf. *Cronologia*,
100). Righini identifies the *domina* with the Iulia Ti. f. Procula who figures prominently in the inscriptions
from a family tomb discovered at Isola Sacra and dated to the Trajanic period; but the name is remarkably
undistinctive, and there is little to suggest that the brickyard owner, whose property evidently passed into
the hands of a Flavia Procula, was identical with the Ostian woman, who has no recognizable connections
with the Flavian *gens* (see Righini, pp. 133–134, with references).

64.   Sig. 10.1; orb. 6.0, litt. 1.2, 1.1–1.2; lin. 1, 1, ?; fr. 16+, 12+, 4.3. – (27068, 1277).

CIL XV 664a   ☽   ➤➤ Σ OSIMI·L·IVLI·RVFI ◀◀
                                              ★ VICCIANA ★
                              caduceus alatus ds. iacens

Vicciana (*sc.* tegula) Zosimi L. Iuli Rufi.

Cf. Steinby, *AIRF* VI, p. 188 no. B 60 (Tav. LIX, 1) = *CIL* XV 664c (= *LSO* 570). For the slave-
officinator Zosimus and his master L. Iulius Rufus, *dominus* of the *figlinae Viccianae* during the
Vespasianic era, see the comment at **62**. The stamp has not been photographed previously.

65.   Sig. ± 8.2; orb. 4.5; litt. 1.2; lin. –, 1; fr. 22+, 18+, 3.4. – (1284).

CIL XV 712   ☽   ★ C̣ ỊVḶI FORTVNATI ★

C. Iuli Fortunati.

Cf. Dressel *ad loc.*: "Hoc sigillum signaculis diversis impressum extat." This example evidently
came from the same die as the copy published at *LSO* 611, where the number of auxiliary lines was
uncertain. According to Helen, the *officinator* C. Iulius Fortunatus who appears here and in a stamp of the
emperor Trajan from the first decade of the second century (*CIL* XV 711) is the same as the Iulius
Fortunatus who worked during the 120's at the brickyards of Seia Isaurica (*Organization*, 143; cf. **101** = *CIL*
XV 1423a); but Steinby points out that the latter might equally well be identified with the M. Iulius
Fortunatus known from a stamp of the Hadrianic age (*CIL* XV 601). She suggests that *CIL* XV 712, which
is similar in type to the Trajanic stamp (711), probably belongs to the earlier period (*Cronologia*, 76 n. 4).

66.   Sig. 9.7; orb. 4.2; litt. 1.2, 0.9–1.1; lin. 1, 2, –; fr. 13+, 9+, 4.4. – (1286).

CIL XV 727   ☽   EX·[PRA· FA]VS·AVG·OP·DOL·CAL
                              CRESCENTIS
                              🌲 nux pinea 🌲

Ex [pra(edis) Fa]us(tinae) Aug(ustae) op(us) dol(iare) Cal(vi) Crescentis.

For the attribution of this stamp to the *figlinae* (*Domitianae*) *Novae*, where the *officinator* Calvius
Crescens worked first under the younger Faustina (Augusta 146), then under her husband, the emperor
Marcus Aurelius, before transferring in the time of Commodus or Septimius Severus to the *figlinae*

Domitianae Minores, see Dressel's comment in *CIL* and Steinby, *Cronologia*, 38 and n. 9, 40 and n. 3. The stamp has not been photographed previously.

67.   Sig. ± 9.6; orb. 4.0; litt. 1.2–1.3, 0.9–1.0; lin. 1, 2, 2; fr. 22+, 18+, 2.3–2.6. – (27080, 1289).

CIL XV 737 (*)      [EX·PRAEDIS·] L·VERI·AVG·OP·DO
                                              LIA·C[·N]VNIDI·FELIC
                                           caduceus ♣

[Ex praedis] L. Veri Aug(usti) op(us) dolia(re) C. [N]unidi Felic(is).

= *LSO* 633. This example, though fragmentary, is slightly better preserved than the copy published in *LSO*, where the marks of punctuation not reported in *CIL* are noted; the triangular mark on this example after the V in VERI seems due to an imperfection in the surface of the brick rather than a mistake of the carver (see the photograph). The *officinator* C. Nunidius (= Nunnidius) Felix had worked at the brickyards of Lucius Aelius already before the latter became the emperor Lucius Verus (see Bloch's comment at *S* 216; on the Nunnidii family of *officinatores*, see Helen, *Organization*, 123–124). For the possible attribution of the stamps of Lucius Verus to the *figlinae Marcianae*, see Steinby, *Cronologia*, 66.

68.   Sig. 9.5; orb. 3.0; litt. 1.6; lin. 1, 2; fr. 27+, 22+, 2.4–3.1. – (1291, 27081, 27082, 27083, 27084, 27085).

CIL XV 748   SEX·PLO·TITIA·EX·PR·C·N̄
                                          ♣

Sex. Plo(ti) Titia(ni) ex pr(aedis) C(aesaris) n(ostri).

The *officinator* Sex. Plo(tius) Titia(nus) is known only from this stamp, which has not been photographed previously. After having seen the photograph, Steinby suggests that Bloch's dating of the stamp to the time of Antoninus Pius or Marcus Aurelius is perhaps too late and should be modified to the time of Hadrian or Antoninus Pius (cf. Bloch, *Indices*, 80).

69.   Sig. 4.5, 1.8 (marg. vid.); litt. A 1.3, V 1.5; fr. 9+, 14+, 4.4. – (1724).

Novum *CIL* XV 776/7   A V   litt. cavis

A(   ) V(   ) *vel* A(   ) U(   ) ?

According to Steinby, this type of stamp (two letters, *litteris cavis*) is Trajanic: cf. *MemAccLinc* 17 (1974), p. 104 no. 19. Although stamps of similar type are frequently found on modern bricks used in the restorations at Ostia, the composition of this fragment is typical of ancient Roman brick in the color of the clay and in the striations and type of inclusions in the fabric. Only one person with these initials is known from stamps of the Trajanic age—the *dominus* M. Annius Verus—but here the letters probably stand for the name of an unknown *officinator* (on Annius Verus, see the comment at **41**). Stamps with similar texts suggest that the letters are more likely to represent a *gentilicium* and cognomen than a single name, Av(   ) or Au(   ) (cf., for example, *CIL* XV 2411; *CIL* X 8042, 119–121; and especially the stamp published by Steinby, *loc. cit.*).

70.   Sig. 10.2; orb. 4.2; litt. 1.1–1.2, 1.1, (O 1.1, D 1.3); lin. 1, 2, 2; fr. 14+, 14+, 3.0. – (93922, 1298).

CIL XV 780b II vel III (*)   ☿   EX PR·AEDIS M AEMILI   sic
                                                  PROCLI·
                                              O ♠ D

Ex pr{·}aedis M. Aemili Procli o(pus) d(oliare).

Cf. Righini, p. 167 no. 130 = *CIL* XV 780b I; *SPASR* I, pp. 38–39 no. 142 = *CIL* XV 780b II ?; *LSO* 656 = *CIL* XV 780c. The mark of punctuation after PR in PRAEDIS is clearly a mistake of the carver (see the photograph). Dressel recognized at least two varieties of *CIL* XV 780b which differed from each other in the orientation of the *signum*: cf. copies 4 and 8, which show the palm frond pointing downward, as it does on both examples in the Michigan collection. On the example published by Righini, and on most of the examples known to Dressel, the palm frond stands upright. A copy from the Aurelian Wall (*SPASR* I) that is said not to show the unusual shape of the A's noted in *CIL* and which can be seen in the photograph published here may have come from yet another die of 780b. M. Aemilius Proclus appears as *dominus* only in the variants of *CIL* XV 780, which Steinby has dated on the basis of form to the time of Antoninus Pius or Marcus Aurelius (cf. also Setälä, *Domini*, 50). According to Dressel, Proclus is to be identified with the *officinator* M. A( ) Pro( ) who worked at the brickyards of Domitia P. f. Lucilla in the years 136–137 (cf. *CIL* XV 1056–1057). But even if homonymy could be established for the *officinator* and the *dominus*, the name is remarkably undistinctive, and it would be unlikely that the stamps refer to the same person: of the 150 private *domini* and the 355 *officinatores* that can be clearly distinguished in second-century stamps, only one man, Vismatius Successus, is known to have filled both capacities (on Vismatius Successus, see Steinby, *MemAccLinc* 17 (1974), 98–100; Helen, *Organization*, 91–92, 131 n. 2; and, most recently, *Arctos* 15 (1981), 20–21; see also Steinby's remarks at *CIL* XV *A* 99). Of course *domini* like M. Aemilius Proclus in whose stamps no *officinatores* appear may well have been brick-producers as well as landowners (cf. Helen, *Organization*, 130).

71.   Sig. 10.2; litt. 1.4, 1.1–1.2; lin. 1, 2, 1; fr. 30+, 26+, 5.4. – (1396).

CIL XV 814 compl. (*)   O   CL·IONICI·CAESARIS·AVGVSTI·L·
                                        ANTIMACHI·
                                        capricornus ds.

Antimachi Cl(audi) Ionici Caesaris Augusti l(iberti).

Dressel's reading and his description of the *signum* are confirmed. It is not entirely certain in what order the lines of text were meant to be read, but the carver probably intended the stamp to be understood as recording only a single name, that of Antimachus, the slave of Claudius Ionicus (cf. Helen, *Organization*, 33–35, 92). For the interpretation of the *signum* as a sign of the zodiac, see Cozzo, 275 and Steinby's confirmation at *Cronologia*, 21 and n. 2 (cf. *Cronologia*, 42 n. 1). Both the slave Antimachus and the imperial freedman Claudius Ionicus are known only from this stamp, which Dressel dated on the basis of paleography and form to the middle of the first century. After having seen the photograph, Steinby is convinced that Dressel's dating is too early and that the stamp belongs rather to the end of the first or the beginning of the second century. According to P. R. C. Weaver, the form of the status indication "Caesaris Augusti l." is otherwise unattested after the time of Tiberius, but the new dating of this stamp lends some support to H. Chantraine's restoration of the formula in a fragmentary inscription which he assigns to the Trajanic period (see Weaver, *FC*, 50 and nn. 1 and 2).

72.   Sig. 10.4+, 3.3; litt. 2.4 (NT 2.5); fr. 16+, 9+, 4.6. – (1304).

CIL XV 824   ▭   APOL·ANT [·L·S]

Apol(loni) Ant(oni) [L. s(ervi)].

The drawing in Lugli, p. 584 no. 3 shows the entire stamp, which has not been photographed previously. Lugli dated the stamp on the basis of paleography and form to the first half of the first century, Blake to near the middle of the first century, but Dressel and Steinby are probably right in assigning it to the end of the Republic or the beginning of the Empire (cf. Lugli, *loc. cit.*; Blake II, p. 56 n. 19; Dressel *ad loc.*). The same slave recurs in two stamps of an M. Antonius, one of which (*CIL* XV *S* 447) gives his full name, Apollonius; the other (*CIL* XV *S* 446) differs from this stamp only in the *praenomen* and in the unabbreviated form of the *gentilicium*. Bloch has suggested that the two Antonii may have been brothers (comment at *S* 447).

73.   Sig. 8.9; orb. 4.4; litt. 1.3–1.4 (TI 1.8), 1.1–1.2; lin. 2, 2, 2; fr. 16+, 17+, 3.4. – (1306, 27089).

CIL XV 842b (*)   ☺   ⇉ Q·ARTICVLEI·PAETI ⇇
                                        SAGITTA·S·F·

Q. Articulei Paeti Sagitta s(ervus) f(ecit).

= *LSO* 701 corr. (*). The Kelsey examples evidently came from the same die as the copy published in *LSO*, where both auxiliary lines can be seen in the photograph above the last two letters of the innermost line (Tav. CXXV). The version in *LSO* preserves the punctuation not reported in *CIL* except the mark after ARTICVLEI, which Marini noted at *Iscr. dol.* 611. For chronological reasons, the Q. Articuleius mentioned here and in other stamps datable to the first decade of the second century is generally identified with Paetus, consul in 101 (= *PIR*² A 1177), rather than his son Paetinus (= *PIR*² A 1173), who probably appears in stamps only in the consular-dating formula of the year 123 (see Bloch, *BL*, 78; Setälä, *Domini*, 69–70). That the slave Sagitta used two arrows as personal emblems in *CIL* XV 842a in place of the palm fronds reported here supports the view that he was acting as *officinator* in his master's service and that Paetus was a *dominus* in the brick industry, as indeed his senatorial status would seem to suggest (cf. Setälä, *Domini*, 70; Bloch, *BL*, 337).

74.   Sig. 9.7; orb. 3.2; litt. 1.6, 1.0; lin. 1, 2, 2; fr. 21+, 10+, 2.6. – (1308).

CIL XV 858 (*)   ☺   ASINIA F MAR[CELL·F]L·
                                        CORINT·O·D· 🌿
                                                EC   v. 3 linea recta   sic

Asinia f(ilia) Mar[cell(i) F]l(avius) Corint(hus) o(pus) d(oliare) <f>ec(it) ?.

= *LSO* 709 corr. (*). This example preserves the innermost series of auxiliary lines, both of which can be seen also in the photograph in *LSO* (Tav. CXXVI), and the mark of punctuation at the end of the first line, which Steinby noted on an example from Santa Maria Maggiore (*RendPontAcc* 46 (1973–1974), 109; cf. also Marini, *Iscr. dol.* 613, where punctuation is printed in the first line after ASINIAE and FI (sic), but not after O in the second line). The oblique stroke extending off the apex of the second A in ASINIA is evidently due to a defect in the die (see the photograph). For the interpretation of the text, see the comments at *LSO* 709 and *CIL* XV 858. The *domina* is to be identified with Asinia Quadratilla, the daughter of Q.

Asinius Marcellus, suffeçt consul sometime in the years 125–130 ca. and *dominus* of the *figlinae Med*(   )
from the early years of Hadrian at least until the year 134 (cf. Setälä, *Domini*, 72–74 with N. Purcell, *JRS* 71
(1981), 215; see also the comment at *LSO* 322). Quadratilla presumably inherited the brickyards of her
father sometime before the year 141, but none of her stamps can be attributed with certainty to the *figlinae
Med*(   ) (cf. Steinby, *Cronologia*, 66). Steinby has pointed out clear connections between the stamps of
Quadratilla and those of M. Flavius Aper which suggest that the two were related to each other, or at least
that they collaborated in managing their properties (see *Cronologia*, 77 n. 3 and, on Flavius Aper, the
comment at **85**). For the career of the *officinator* Flavius Corinthus and the dating of this stamp to the early
years of Antoninus Pius, see Steinby, *Cronologia*, 66–67 n. 11, 78 n. 4.

75.   Sig. 10.8; orb. 4.5; litt. 1.5–1.6, 1.7; lin. 1, 1; fr. 18+, 14+, 3.0–3.7. – (1310).

CIL XV 887b   ☽   L·VELLICI·SOLLERTIS
                                          EX·F   v. 2 linea recta

        Ex f(iglinis) L. Vellici Sollertis.

        Cf. *LSO* 721–724 = *CIL* XV 887a,c; *S* 241 corr.; 888. The publication of this example completes
the photographic record of the known stamps of the consular *dominus* L. Bellicius Sollers, on whom see
Setälä, *Domini*, 76–78. Bloch and Steinby agree in dating his stamps to the last years of Trajan or the first
years of Hadrian: by the year 123 his brickyards had already passed into the hands of his wife, Claudia
Marcellina (cf. Bloch, *BL*, 180; Setälä, *loc. cit.*).

76.   Sig. ± 8.6; orb. 4.0; litt. 1.4, 1.2–1.5; lin. ?, ?; fr. 14+, 10+, 3.7. – (1311).

CIL XV 904a   ☽   ⋙ HERMETIS ⋘   vers. falc.
                                    ★ C CAL[PETA·FAVOR ★]

        Hermetis C. Cal[peta(ni) Favor(is)].

        = *LSO* 731; cf. Righini, pp. 174–175 no. 140 (Tav. XXV, fig. 2) = *CIL* XV 904d (= ? Camilli,
RendAccLinc 28 (1973), p. 309 no. 22); *LSO* 732 = 904e corr.; *CIL* XV *A* 35 = 904 var. The photograph of
the Kelsey example shows more clearly a portion of the stamp preserved in its entirety in Cozzo, Tav. VII,
fig. 17. The transcription begins with the uppermost line. During the Trajanic era C. Calpetanus Favor was
the most prominent *officinator* at the *figlinae Marcianae*, a section of which, the *Favorianae*, derived its
name from his cognomen at the end of the second century (cf. Bloch, *BL*, 335–336; Steinby, *Cronologia*, 65.
For the Calpetani, see the comment at **26**; on the *figlinae Marcianae*, see the comment at **28**; on the
Favorianae, see the comment at **25**). The slave Hermes, who was manumitted shortly after this stamp was
made, in the first decade of the second century, continued to work at the *figlinae Marcianae* as *officinator*
under Hadrian (cf. Bloch, *BL*, 48 n. 52; Steinby, *Cronologia*, 65 and n. 6).

77.   Sig. 10.0+, 3.9; litt. 1.3–1.4, 1.3–1.4; fr. 14+, 13+, 5.9. – (1312).

CIL XV 910a   ▭   M CALPVRNI
◄◄◄◄◄◄◄◄◄◄
[P]HRONIMI

M. Calpurni [P]hronimi.

= *CIL* XV 2083 ? (cf. Bloch, *Indices*, 22). M. Calpurnius Phronimus is known only from the several variants of *CIL* XV 910, of which a and b can be dated from discoveries *in situ* to the beginning of the second century (cf. Bloch, *BL*, 79; Blake II, p. 131 n. 75). The stamp has not been photographed previously.

78.   Sig. 10.4+; litt. 2.1–2.3; fr. 10+, 7+, 3.3. – (1314).

CIL XV 931b   [E]X F CLAV IV[LL]   litt. cavis

[E]x f(iglinis) Clau(di) Iu[ll(i)].

According to Steinby, this type of stamp (impressed letters greater than 2 cm. in height) is found exclusively on *bessales* of the Trajanic or early Hadrianic period (*Cronologia*, 19). For chronological reasons, therefore, the *dominus* Clau(dius) Iull(us) should probably not be identified with the Ti. Cl(audius) Iul( ) known from a fragmentary stamp approximately datable to the late Hadrianic or early Antonine period, as Setälä rightly notes (*Domini*, 90; cf. Coste, *RendPontAcc* 43 (1970–1971), p. 96 no. 4 (fig. 7)); but whether he can be identified with a homonymous legate under the governor of Asia, as she proposes, remains doubtful for the same reasons, since the tenure of the office has been dated, albeit tentatively, to the time of Vespasian (cf. *Domini*, 91). In all probability, the *dominus* Clau(dius) Iull(us) is known only from the two variants of *CIL* XV 931, neither of which has been photographed previously.

79.   Sig. 10.6, 3.7; litt. 1.6, 1.4; fr. 17+, 19+, 2.3–2.8. – (1315).

CIL XV 971 (*)   ▭   ·CVSPI·
MELICHRYSI

Cuspi Melichrysi.

The triangular mark of punctuation after CVSPI is faint but clear; the one before is less secure; the final I of MELICHRYSI is only faintly visible (see the photograph). Cuspius Melichrysus, perhaps a freedman of the Q. Cuspius who appears in *CIL* XV 969, is known only from this stamp, which Steinby has dated to the Julio-Claudian period after Augustus (*AIRF* VI, p. 194 n. 2; cf. p. 173 no. B 7 = *CIL* XV 971. For *CIL* XV 969 (= *LSO* 761), "tardo-repubblicano od augusteo", see p. 172 no. B 1; cf. also *CIL* XV 970, S 263, and 972—all clearly of the same group). A fragmentary rectangular stamp MELIC[recently published by Steinby, *Largo Argentina*, p. 331 no. 146 (Tav. LXXIX, fig. 6) perhaps belongs to the same Cuspius Melichrysus before his manumission, as Steinby suggests in her comment.

80.   Sig. 8.5; orb. 4.1; litt. 0.9–1.1, 1.0; lin. 1, 1, 2; fr. 18+, 15+, 4.1. – (1318).

CIL XV 1002   ☽   ⫷ AGATHOBVLI ⫸
                                            DOMITI·TVLL
                                               ⫷

Agathobuli Domiti Tull(i).

   Although this example is marred by encrustation, it preserves the entire stamp, which has not been photographed previously (cf. Righini, p. 186 no. 144). Agathobulus, the slave first of Cn. Domitius Tullus, then of Domitia Cn. f. Lucilla, worked at the brickyards of the Domitii for nearly thirty years, before and after his manumission around the year 114. This stamp, his earliest, can be dated to shortly after the year 93/4, when the death of Tullus' brother Lucanus dissolved their partnership in the brick industry (on Agathobulus, see Steinby, *Cronologia*, 51 and n. 6, 52, 55; for the *societas* of the Domitii, see Helen, *Organization*, 114, and, further on Cn. Domitius Tullus, Setälä, *Domini*, 36–37; for the *gens Domitia*, see the comment at **61**).

81.   Sig. ± 9.6; orb. 4.5; litt. 1.2, 1.1; lin. 1, 2, 2; fr. 19+, 13+, 2.9. – (1322).

CIL XV 1034 (*)   ☽   FAVSTVS DOMITIAE·P·F·LVCILL
                                        PAET·ET·APRONIA·COS   a. 123
                                       nux pinea foliis circumdata

   Faustus Domitiae P. f. Lucill(ae) Paet(ino) et Apronia(no) cos.

   For a better preserved, but fragmentary example, see Cozzo, Tav. XLVI, fig. 153. The slave Faustus had worked as *officinator* at the brickyards of the Domitii since they belonged to Cn. Domitius Tullus († 106/7). This stamp, his last, is among the earliest of the younger Domitia Lucilla, who sometime in the first years of Hadrian inherited from her mother the *figlinae Caninianae* and *Domitianae* (for the dates see Bloch, *BL*, 46, 320 n. 256; on Faustus, see Steinby, *Cronologia*, 51, 52, and *CIL* XV *A* 144; for Domitia P. f. Lucilla and the *gens Domitia*, see the comment at **61**).

82.   Sig. 7.7; orb. 3.6; litt. 0.9–1.3, 1.2–1.4; lin. ?, 2, 1; fr. 16+, 18+, 4.8. – (1337, 27098, 27097).

CIL XV 1102a   ☽   CN·DOMITI   vers. falc.
                                   ⫸ CLEMENTIS ⫷

   Cn. Domiti Clementis.

   Cf. *LSO* 866 = *CIL* XV 1102b. For the auxiliary lines of 1102a (1, 2, 1), see Steinby, *AIRF* VI, p. 175 no. B 15 and the photograph in Van Essen, *Santa Prisca*, Pl. LXX, 3, which is mistakenly identified in the catalogue as 1102b (p. 250 no. 34–8/34). The transcription of the text begins with the uppermost line. The stamp has been found in the Flavian palace on the Palatine and can therefore be dated to the period of Domitian's reign before the year 92 (see Bloch, *BL*, 28; cf. p. 41). It is perhaps the earliest of the freedman-*officinator* Cn. Domitius Clemens, who recurs in two brick stamps of the Trajanic age (*CIL* XV 1102b, *LSO* 867 = *S* 286 compl. (*)) and a *dolium* stamp of uncertain date (*S* 502; cf. Steinby, *Cronologia*, 56 and n. 9). For the *gens Domitia,* see the comment at **61**.

83.   Sig. 11.7+, 2.3; litt. 1.9–2.0; fr. 20+, 17+, 2.9–3.3. – (1344).

CIL XV 1121c II   ▭   L·DOMITI

L. Domiti.

The Kelsey example has the same dimensions as the copy of *CIL* XV 1121c in the Di Bagno collection (no. 15), shows the same squared shape of the M, and probably came from the same die (see the photograph; the comment in *CIL*; and Righini, p. 207 no. 171). The two examples published at *LSO* 1223 and De Rossi, *Tellenae*, p. 96, fig. 216 (cf. p. 98 no. 55,3) are both fragmentary, but they seem to be copies of the same stamp, which, then, must be a version of *CIL* XV 1121c impressed from one of the different dies reported by Dressel (= *CIL* XV 1121c I; cf. also Descemet, *Inscr. dol.,* p. 224 no. 66: *litt.* 1.6 cm. (= 1121c I ?);*LSO* 886 = *CIL* XV 1121a I; *LSO* 887 = *CIL* XV 1121a II or 1121c III ?). L. Domitius is known only from the several variants of *CIL* XV 1121, of which a and b have been found in the imperial galleys from lake Nemi and can therefore be dated to the time of Caligula (cf. Gatti, *Navi*, 345–346); to judge from the shapes of the letters, particularly the M, the variety of 1121c in the Michigan collection belongs to a slightly later period, probably the age of Claudius.

84.   Sig. ± 7.4; litt. 1.5–1.6; lin. ?, ?; fr. 14+, 10+, 3.2 – (1736).

CIL XV 1139   [O]   FELICIS FLA[VIAES DO]MITIL
                                    ramus palmae corona vinctus

Felicis Fla[viaes Do]mitil(lae).

Descemet's description of the *signum*, rejected by Dressel at *CIL* XV 1139, 3, has since been vindicated by Steinby: see *MemAccLinc* 17 (1974), 93. Of the three known Flaviae Domitillae of successive generations, the *domina* is generally identified with the youngest (= *PIR²* F 418), grand-daughter of the emperor Vespasian and wife of T. Flavius Clemens, consul in 95, because *CIL* XV 1139 can be dated from discoveries *in situ* to the Domitianic age (see the places of discovery reported in *CIL*, especially no. 2). The same Flavia Domitilla appears as the owner of a slave Secundus in a contemporary stamp recently published by Steinby, who suggests that Domitilla's brickyards may have been located on the Via Ardeatina, at Tor Marancia, where she evidently owned property (see Steinby, *MemAccLinc* 17 (1974), p. 103 no. 14 (Tav. V, fig. 28); cf. Setälä, *Domini*, 38). Since the stamps of Domitilla belong to the Domitianic period, the *officinator* Felix should not be confused with the T. Fl(avius) Felix whose only stamp was consular-dated in the year 123 (cf. *CIL* XV 1255).

85.   Sig. 9.0; orb; 3.6; litt. 1.2–1.4, 0.9–1.0; lin. 1, 3, 2; bip. 60, 60, 3.6–4.1 – (1130, 1060).

CIL XV 1146b compl. (*)   ◐   ·EX·PR·FLAVI APRI OPVS·
                                                DOLIA LARCIO

Ex pr(aedis) Flavi Apri opus dolia(re) Larcio(nis ?).

The die was cut with the upper parts of the letters toward the outside; the marks of punctuation are round. The shape of the L's is unusual, but not without parallel in stamps: cf., for example, *CIL* XI 6689, 320; 6691, 30 (see the photograph). Both examples from the De Criscio collection preserve in its entirety a stamp that Dressel knew only in imperfect condition (see the Introduction, p. 6). Neither seems to be identical with two other copies owned by De Criscio, which Mommsen included in the *additamenta* to *CIL* X and which Mau described as having a tall I in LARCIO; but the circumstances of discovery reported in *CIL* and the description of the bricks as *tegulae magnae quadratae* are suspiciously consistent with the examples in the Michigan collection, both *bipedales,* and the two copies now in the Kelsey Museum may in fact be the same ones that Mau published at *CIL* X 8331, 1 (where, however, the mark of punctuation at the end of the first line is omitted. As Kelsey's diaries make clear, Mau played an important role in securing the Italian government's permission to export certain of De Criscio's inscriptions to the United States: see the Introduction, p. 5).

Despite the reservations of Dressel, who thought that 1146b might have been an African version modeled after the Roman original, 1146a, the form (orbicular) and the formula (*opus doliare*) are peculiar to urban stamps, and, as Mommsen realized, both variants undoubtedly came from Rome (see the comments at *CIL* XV 1146 and *CIL* X 8331, 1; for the criteria used to distinguish urban stamps, see Steinby, *Cronologia*, 13, and, for their geographical distribution, *RE* Supplb. XV, 1492–1493 (cf. *Appendice,* p. 55 n. 3); see also Bloch's comment at *S* 460). The *officinator* Larcio (?) is known only from these stamps, which can be approximately dated to the middle of the second century, shortly after M. Flavius Aper had inherited from (his mother ?) Flavia Seia Isaurica the *figlinae Fabianae, Publilianae,* and *Tonneianae* (cf. Steinby, *Cronologia*, 77 and n. 3; Setälä, *Domini*, 113–115; on Seia Isaurica see the comment at **102**). Since Aper's stamps date from no later than the 160's, Steinby and Setälä are probably right in identifying him with the consul of 130 rather than the consul of 176, as Bloch proposed in his *Indices* (p. 31), reversing his earlier opinion (*BL*, 280. See also the comment at **74**).

86.   Sig. 10.6, 4.5; litt. 1.2; fr. 27+, 16+, 5.3 – (1398

CIL XV *S* 316 compl.   ▭   quindecim semina ?
                        situla   Q·GRA·APOL   sistrum
                        decem cymbala ?

        Q. Gra(ni) Apol( ).

CIL XV *S* 316 is completed in the description of the decoration, which can be seen also in the drawing by Giuseppe Gatti in *BullCom* 29 (1901), p. 133, fig. 2. The interpretation of the figures below the text, which vary in shape from being oval to being nearly circular, is uncertain, but since the other decorative elements clearly pertain to the cult of Isis, they are perhaps more likely to represent cymbals (cf. Steinby, *MemAccLinc* 17 (1974), p. 106 no. 25 (Tav. VI, fig. 36)) than eggs (indeed, it now seems that eggs did not figure in the decoration of brick stamp texts: see the comment at *LSO* 1078 (= *CIL* XV 1501 corr.)). Q. Gra(nius) Apol(lonius ?) or Apol(linaris ?) is known only from this stamp, which can be dated on the basis of paleography and form to the first half of the first century. Possibly a freedman of the Q. Granius who accused Calpurnius Piso of treason in A.D. 24 (= *PIR²* G 207), Apol( ) was clearly connected with the Q. Granius Blastus whose brick stamp has been dated by Steinby to the middle of the first century or shortly before (cf. *MemAccLinc* 17 (1974), p. 103 no. 15. Gatti, who assigns the stamp of Apol( ) to the end of the Republic or the beginning of the Empire, suggests a connection with Q. Granius the *praeco*, a contemporary of L. Licinius Crassus, consul in 95 B.C.: cf. Munzer, *RE* s.v. "Granius" (1912), 1818 no. 8). The stamp of Blastus, like that of Apol( ) shows ornamentation appropriate to the cult of Isis: the text appears *inter duos ramos palmae*. In both cases the decoration should be seen as alluding to the brickyard where the brick originated, the *figlinae ab Isis* or *Isiacae,* which derived their name in the Trajanic period either from their proximity to a temple of the goddess or from their location in the homonymous region of Rome (see Steinby, *Cronologia*, 46 and n. 5; Gatti, *loc. cit.*).

87.   Sig. ?; orb ?; litt. 1.5; lin. 1, ?; fr. 8+, 5+, 4.2 – (1346).

CIL XV 1181a   [☉]   [COSMI] M HER POL [·SER]

[Cosmi] M. Her(enni) Pol(lionis) [ser(vi)].

Datable by type to the beginning of the second century, *CIL* XV 1181a is perhaps the earliest stamp of the *dominus*, who can now be positively identified as M. Annius Herennius Pollio, suffect consul with his father Publius in A.D. 85 (see Setälä, *Domini*, 127–129, where, however, the father and son are confused: cf. *AE* 1975, nos. 21, 131). Of the three *officinatores* known to have worked at Pollio's brickyards, the slave Cosmus and L. Sessius Successus recur in stamps of Flavia Seia Isaurica, who evidently inherited his claylands in the early Hadrianic period (cf. *CIL* XV 1422, *S* 107; for Seia Isaurica, see the comment at **102**). It is tempting to suppose that Cosmus and Successus worked at the same brickyards under Seia Isaurica as they had under Herennius Pollio, in which case the *figlinae* identified in the Hadrianic stamp of Successus as the *Publilianae* are the same ones that Herennius Pollio owned during the Trajanic period; but either or both of the *officinatores* might have transferred to new brickyards when the claylands changed hands, and the evidence is therefore insufficient to support any firm conclusions about the attribution of Pollio's stamps (for the *Publilianae*, see Steinby, *Cronologia*, 75–78).

88.   Sig. ± 9.8; orb. 4.3; litt. 1.3 (S in fine 0.6), 1.2–1.3; lin. 1, ?, 1; fr. 18+, 12+, 3.7–4.2 – (1348).

CIL XV 1203 II   ☻   EX·PEREDIS [C IVLI·APOLL]INARIS
                                            FA CET·M[AG]NIO
                                               ⋘

Ex peredis [C. Iuli Apoll]inaris fa(cit) Cet(ennius) M[ag]nio.

The interpretation given above seems preferable to the alternative, *facet Magnio*, suggested by Dressel: evidently the carver has twice written E for AE—in PEREDIS for PERAEDIS and in CET for CAET (for *Cet(ennius)*, cf. *CIL* XV 942; for *fa(cit)*, see Bloch, *Indices*, 96). The spelling PEREDIS, though anomalous, cannot be dismissed as a scriptural error, since *CIL* XV 1203 was impressed from at least two different dies: an example from the Aurelian wall shows the palm frond pointing to the left and a dash in the second line in place of the triangular mark of punctuation noted here (cf. *SPASR* I, p. 47 no. 181 = Marini, *Iscr. dol.* 950 = *CIL* XV 1203 I). That the *gentilicium* of the *officinator* should be completed Caet(ennius), as Helen had already proposed, has now been confirmed by a recently published correction of *CIL* XV 1377, which moreover adds a new *domina* to the list of his associates (cf. Camilli, *RendAccLinc* 34 (1979), p. 209 no. 80 (Tav. XI, 80); Helen, *Organization*, 23, 141). During the Hadrianic period Caetennius Magnio worked at the brickyards of at least two different owners of *figlinae Antull(ianae)*, to which, therefore, Steinby tentatively assigns *CIL* XV 1203 as well. She suggests that C. Iulius Apollinaris may have owned the *figlinae* before the year 134; he is known only from the variants of 1203, neither of which has been photographed previously (see Steinby, *Cronologia*, 26; cf. Setälä, *Domini* 131 and n. 2).

89.   Sig. 9.6; orb. 4.3; litt. 1.3–1.4, 0.9–1.0, 0.6–0.7; lin. 1, 2, 2; fr. 17+, 10+, 3.9–4.2. – (1349).

Novum *CIL* XV 1204/5   ☻   EX·FIG [C·IVLI DIO]SCORI
                                               ANSO [III ET V]ERO II   sic   a. 161
                                                  CO2

Ex fig(linis) [C. Iuli Dio]scori An⌈t⌉o(nini) [III et V]ero II cos.

The new stamp may have come from the same die as *CIL* XV 1204, from which it seems to differ only in the error in the second line, S for T in ANSO: simple mistakes of the carver could evidently be corrected by recutting the die (see the comment at **32**; cf. Marini, *Iscr. dol.* 523, where punctuation is printed also after FIG, $\overline{\text{III}}$, and ET). In regard to the consular dating of A.D. 161, Dressel cites *CIL* VI 1119, which shows that the brick stamps were not necessarily made before March 7, when Antoninus became the emperor Marcus Aurelius (cf. *CIL* XV 353 of the same year, where the consuls are announced as *Augusti nostri*). As a rule, workmen did not need new stamps before April or May: see Bloch, *CP* 39 (1944), 254–255. The *dominus* C. Iulius Dioscorus is not otherwise known (cf. Setälä, *Domini*, 132 and n. 1).

90.   Sig. 10.5; orb. 4.0; litt. 1.3–1.4, 1.0–1.3, lin. 1, 2, 2; fr. 26+, 27+, 3.3–3.9. – (1350).

CIL XV 1220 (*)   ☺   SEVERO·ET STLOGA [CO]S OPVS FIG   a. 141
                                                  D P·IVLIAE SATVRN
                                                  corona lemniscata

Severo et Stloga [co]s. opus fig(linum) d(e) p(raedis) Iuliae Saturn(inae).

= *LSO* 936 (*). Marini, *Iscr. dol.* 499 shows punctuation between words except between OPVS and FI (sic). Steinby has identified the *domina* Iulia Saturnina, whose stamps were consular-dated in the years 137, 139, 141, and probably 150, as the daughter of C. Iulius Stephanus, *dominus* during the middle years of Hadrian's reign, because the *signum* that appears on this stamp and others of hers was frequently used by Stephanus as a rebus (*Cronologia*, 44 and n. 1; cf. Setälä, *Domini*, 140–141). There is some reason to believe that the stamps of Stephanus belong to the *figlinae Genianae,* and it is quite likely that Saturnina inherited the brickyards of her father sometime after the year 132, but the evidence is insufficient to support the firm attribution of her stamps to that *figlinae* (cf. Steinby, *loc. cit.*; Setälä, *Domini*, 134; the stamps of Saturnina: *CIL* XV 1218–1222; *S* 323 (cf. *LSO* 932–937); *A* 48, which should probably read FICLIN at the end of the first line: cf. Camilli, *RendAccLinc* 34 (1979), p. 208 no. 74 (Tav. XI, 74)). In any case, the *domina* Iulia Saturnina should not be confused with the *officinatrix* of the same name who worked at the *figlinae Oceanae* under the emperor Marcus Aurelius (see Steinby, *Cronologia*, 70 and n. 4 and the comments at *CIL* XV 1221 and *LSO* 937).

91.   Sig. 7.0, 2.5–2.8; litt. 1.6–1.7; sesq. 46, 32+, 3.7. – (1126).

CIL XV 1237   ▭   LEPIDI·

Lepidi.

The stamps of Q. Lepidius date from the early Julio-Claudian period and are of Campanian origin: see the comment at **7** and the Introduction, p. 8 and n. 29. This example from the De Criscio collection preserves the entire stamp, which has not been photographed previously.

92.   Sig. 8.6; orb. 6.4; litt. 0.7–1.0; lin. 1, 2; fr. 15+, 16+, 3.6 – (81.1.1).

CIL XV 1248a   ◯   aquila alis expansis ss. respiciens
                                           duobus cornibus copiae insistit
                                           ★ L ★ LVRI·BLANDI ★

L. Luri Blandi.

Cf. *LSO* 953 = *CIL* XV 1248b. This example is said to have come from the Roman villa on Giannutri, where several copies of *CIL* XV 1248a have already been discovered (cf. *NSc* 1935, 145, 152). The
signum is located inside a circular perimeter, between the cusps of a narrow lunate field on which the text is
impressed. Of the ears of corn and bunches of grapes described by Marini at *Iscr. dol.* 1016, the latter,
perhaps, can be seen overflowing the mouth of one of the cornucopiae (see the photograph; cf. *SPASR* I, p.
46 no. 185). The stamp is identical in form to only two others, one of which has been found in the
Colosseum and can therefore be dated to the Vespasianic age (*CIL* XV 1275b = Cozzo, Tav. XXVII, fig. 86;
cf. *CIL* XV 62 = Cozzo, Tav. XXVII, fig. 87). The rarity of the type has suggested to Steinby that all three
stamps came from the same *figlinae,* the *Marcianae,* and were in use during approximately the same period
(*Cronologia,* 64 and n. 7; but cf. n. 5). It is at least certain that L. Lurius Blandus was active before A.D. 79:
his name was stamped on a *mortarium* found at Pompeii (*CIL* XV *S* 511). On the L. Lurii family of
officinatores, see Helen, *Organization,* 122 (cf. *CIL* XV *A* 106; 169; and 171 with Coste, *RendPontAcc* 43
(1970–71), p. 107 no. 12 (fig. 14)).

93.   Sig. ?; orb. ?; litt. 1.1–1.6; lin. 1, ?; fr. 14+, 10+, 3.1. – (1354).

CIL XV 1284 compl.   ☾   [➤ C MAR]CI PYRRICHI ◀

[C. Mar]ci Pyrrichi.

CIL XV 1284 is completed in the description of the decoration at the end of the line. C. Marcius
Pyrrichus is known only from this stamp, which Steinby has dated on the basis of form to the early Flavian
age (cf. *RendPontAcc* 46 (1973–1974), 112). It is the only orbicular type and is probably among the latest in
an otherwise homogeneous group of stamps belonging to the slaves and freedmen of an unidentified C.
Marcius L. f. (cf. Steinby, *Cronologia,* 15 n. 1; *RE* Supplb. XV, 1518).

94.   Sig. 7.5+, 3.9; litt. 2.0, 2.0; lin. 1, 2, 1; fr. 17+, 10+, 3.2. – (2408).

Novum *CIL* XV 1293/4   ▭   ]MARSID[
                                                            ]ARANTH[

[C.] Marsid[i Am]aranth[i] ?

The H in ligature TH is uncertain: see the photograph. The restoration of the first line is suggested
by two one-name stamps of the C. Marsidii Faustus and Philomusus (*CIL* XV 1294, 1295 = *LSO* 970). I
know of no other C. Marsidii, and the *gentilicium* is rare (cf. Schulze, *LE,* 189; *CIL* XI 4486). The new
stamp, which is similar in type to 1294, probably records the name of a freedman, Amaranthus (or
Amarantus) (cf. Solin, *Beiträge,* 115; no cognomina with this combination of letters are found in Kajanto,
LC). Paleography and form suggest a date in the second half, perhaps the third quarter, of the first century.

95.   Sig. 5.4+, 2.7; litt. 1.2, 0.9–1.1; fr. 20+, 21+, 2.6–3.0. – (1358).

CIL XV 1317   ▭   L·NA[EVI· ⊤⌐
                                                 T·POM[PONI ⌐

L. Na[evi et] T. Pom[poni].

Several freedmen of L. Naevius and T. Pomponius are known from stamps of the first century, but there is no evidence of kinship between the two families, and the partnership that can be inferred from this stamp was probably short-lived: cf. Helen, *Organization,* 113–115 (stamps of the freedmen: *CIL* XV 1332; *S* 345; *S* 357, which should read T·POMPONI·PRIM: cf. *NSc* 1917, 175; Camilli, *RendAccLinc* 34 (1979), p. 209 no. 79 (Tav. XI, 79) = *S* 357 var.; cf. *CIL* XV 1316). For the attribution of all undated stamps of the various Naevii to the *figlinae Naevianae* of the Augustan or early Tiberian period, see Steinby, *Cronologia,* 67 and nn. 7 and 8, 131; for the consular-dated Republican stamps of the L. Naevii, see Bloch, *AJA* 63 (1959), 235 n. 40; and, further on the Naevii, Blake I, 299. The stamp has not been photographed previously.

96.   Sig. 7.6+, 2.7–2.8; litt. 1.7–2.1; fr. 17+, 29+, 3.2–3.6. – (1359).

CIL XV 1323a   ▭   C·ИAEVI·[ΛSC]

C. Naevi [Asc(lepiadis)].

Cf. Camilli, *RendAccLinc* 28 (1973), p. 309 no. 23 (Tav. IV, fig. 3) = *CIL* XV 1323a var. This example, in which the letters are uneven and the interpuncts round, was evidently carved by the same man who made *CIL* XV 1316b (= *LSO* 984) and the variant of *CIL* XV 1318 published by Camilli, *RendAccLinc* 34 (1979), p. 208 no. 77 (Tav. XI, 77): note in particular the shape of the N's, but also the A's and the V's. The stamp has been found in the House of Livia on the Palatine and can therefore be dated to at least as early as the Augustan period (cf. *NSc* 1967, 313); the character of the lettering suggests an even earlier date, at the end of the Republic (cf. the comment on the preceding stamp). C. Naevius Asclepiades is not otherwise known.

97.   Sig. 10.6+, 2.4–2.5; litt. 1.7–1.8; fr. 17+, 10+, 4.0. – (1721).

CIL XV *S* 342 corr.   ▭   C·NAE·BAL signum incertum

C. Nae(vi) Bal(bi ?).

= *CIL* XV *A* 59; cf. *CIL* XV 1324 = 1973 = *S* 343 = *LSO* 986; *CIL* XV 1974; *CIL* XV 1975 = 1324 var. = *A* 58. The horizontal stroke of the L in ligature AL is oblique; the two loops of B are detached; the interpuncts are round (for the shape of the B, see the comment at **98**). Bloch knew the stamp only from the version transcribed by Ashby in his copy of *CIL* XV,1 and published by him in *PBSR* 5 (1910), 292, where the ligature in the *gentilicium* comprises only the letters AE. On this example the N appears in ligature NAE, as it does on both copies reported by Steinby in the Museo Comunale at Velletri (cf. *CIL* XV *A* 59), and the *signum* tentatively identified by Ashby as an amphora more closely resembles a caduceus (see the photograph). The stamps of C. Naevius Bal(bus ?) can be approximately dated to the time of Augustus: several copies of *CIL* XV 1324 have been found in the House of Livia on the Palatine (cf. Ashby, *loc. cit.*; *NSc* 1957, 80, 91, 94; see also the comment at **65**. For the cognomen Balbus, cf. Munzer, *RE* s.v. "Naevius" (1935), 1562–1563 nos. 10–11).

98.   Sig. 7.0+, 2.6; litt. 2.0 (O 1.6); fr. 12+, 11+, 3.0. – (1727).

CIL XV 1333 var. 2   ▭   [C·NAEVI·P]HILOMV

[C. Naevi P]hilomu(si).

= Marini, *Iscr. dol.* 1070a; cf. Camilli, *RendAccLinc* 34 (1979), p. 209 no. 78 (Tav. XI, 78) = *CIL* XV 1333 var. 1. The Kelsey example differs from *CIL* XV 1333 in the lack of a final ligature VS and from the recently published variant in the shape of the letters: compare the photographs with the drawings of *CIL* XV 1333 in Lugli, p. 554, fig. 2 and *NSc* 1957, p. 91, fig. 19a. The other variant of 1333, [NA]EVI·BHILOM[ ], was perhaps carved by the same man who made *CIL* XV 1975 (= *A* 58) and *CIL* XV *A* 59 (= **97**) for C. Naevius Bal(bus ?): note the shape of the B's, in which the upper and lower loops meet only at the vertical stroke (a stamp of Philomus(us) occurs with one of Bal(bus ?) in the House of Livia on the Palatine: cf. *NSc*, 1957, 91). C. Naevius Philomus(us) is known only from the three variants of *CIL* XV 1333, all of which are approximately datable to the Augustan period (see the comment at **95**).

99.   Sig. 6.3; litt. O 3.0, V 3.6 ?; fr. 16+, 8+, 3.1. – (27127).

Novum *CIL* XV 1336/7   O V   litt. extantibus

     O( ) V( ) *vel* O( ) U( ) ?

     Perhaps the serif on the second letter stands for an S in ligature V̲S̲ (see the photograph and the comment on the preceding stamp), but the text probably records only two initials representing the *gentilicium* and cognomen of an unknown *officinator* (see the comment at **69**; cf. *LSO* 1183). It is just possible that the stamp is here presented upside down and that the text should be transcribed Λ O, but the position of the serif and the lack of a transverse stroke on the V recommend the reading given above. None of the usual criteria has yet provided a reliable means of dating this type of stamp, which Steinby suggests might belong to either the Trajanic or the Severan period.

100.   Sig. 10.4+, 3.4; litt. 1.2, 1.3; fr. 15+, 11+, 3.4. – (1365).

CIL XV 1356 II ?   ▭   Q·PAPIRI
                                          FIGVL[I 🌿]

     Q. Papiri Figul[i].

     = *LSO* 1005; cf. the drawing in Lugli, p. 554, fig. 5 (= *CIL* XV 1356 I ?). The Kelsey example apparently came from the same die as the fragmentary copies discovered at Ostia; the copy reproduced by Lugli seems to differ from these examples in the shapes of the F and the second P and in the disposition of the text at the end of the second line, but the discrepancies are subtle and may be due to a slight distortion in Lugli's illustration rather than a genuine variance in the paleography of two different stamps. Nonetheless, the existence of various dies of *CIL* XV 1356 was reported by Dressel, who dated the stamps on the basis of paleography and form to the end of the first or the beginning of the second century. Q. Papirius Figulus is otherwise known only from *CIL* XV 1357, where he appears as Q. Figulus (for the cognomen Figulus, see Marini's comment at *Iscr. dol.* 858 and Kajanto, *LC*, 322).

101.   Sig. 9.5; orb. 4.5; litt. 1.4–1.5, 1.4–1.5, 0.8–1.0; lin. ?, 2, 2; fr. 23+, 13+, 4.9–5.4. – (1367).

CIL XV 1414   ☺   M VINIC FORTVNAT EX P Q SC̣
                                          P P APR ET PA̰ET   a. 123
                                                   COS   v. 3 linea recta

     M. Vinic(i) Fortunat(i) ex p(raedis) Q. Sc( ) P( ) P( ) Apr(oniano) et Paet(ino) cos.

The *officinator* M. Vinicius Fortunatus recurs in a contemporary stamp of the *figlinae Astivianae* (*CIL* XV 13), to which Steinby therefore assigns *CIL* XV 1414 as well (*Cronologia*, 26 and n. 9; see also the comment at **51**). Her interpretation of the name of the *dominus*, Q. Sc( ) P( ) P( ), seems preferable to Dressel's suggestion, Q. Sc( ) P( ) p(ater ?), which is without parallel in brick stamps (but cf. *CIL* XV 1307, 2199).

102.   Sig. 9.4; orb. 4.7; litt. 1.2, 1.2; lin. 1, ?, 1; fr. 13+, 13+, 4.3. – (1369).

CIL XV 1423a   ☾   IVLI FORTVNATI DE PRAEDIS
                                                SEIAE ISAVRICAE
                                         nux pinea

Iuli Fortunati de praedis Seiae Isauricae.

Cf. Righini, p. 228 no. 199 = *CIL* XV 1423b compl.; 1423a has not been photographed previously. The *officinator* Iulius Fortunatus cannot be positively identified, nor can it be determined to which of the six *figlinae* owned by Flavia Seia Isaurica this stamp belongs (see the comment at **65** and, for Seia Isaurica, Setälä, *Domini*, 119–121; cf. Helen, *Organization*, 57). Since the stamp can be dated from discoveries *in situ* at Hadrian's Villa to the beginning of the 120's, it is not likely to have come from the *figlinae Tonneianae*, which Isaurica seems not to have owned until the 140's, or the *Tur(rianae)*, which are unattested before the year 134, but there is little to choose between the *figlinae Aristianae, Caelianae, Publilianae,* and *Tonneianae,* all of which belonged to Isaurica during the early years of Hadrian's reign (for the date of 1423a, see Bloch, *BL*, 155; for the brickyards of Seia Isaurica, see Steinby, *Cronologia*, 26, 30, 40–41, 76, 99, 100).

103.   Sig. 7.3; litt. 1.4–1.5 (TI 1.6); lin. 1, 1; fr. 24+, 14+, 3.2. – (1373).

CIL XV 1442   O   L·SERVI·FORTVNATI·
                                  ramus maior inter duos ramos minores

L. Servi *vel* Servi(li) Fortunati.

The brick is broken and mended. The Fortunatus who appears in this stamp, which can be approximately dated on the basis of paleography and form to the Domitianic age, does not appear elsewhere and should not be confused with the *officinator* (L. ?) Servilius Fortunatus who worked at the *figlinae Macedonianae* and perhaps also at the *Terentianae* at least half a century later (cf. the comment of Dressel, who dated the stamp to the middle of the first century; for Servilius Fortunatus, see Steinby, *Cronologia*, 60 and n. 3). The stamp has not been photographed previously.

104.   Sig. 9.3; orb. 4.3; litt. 1.2, 1.2; lin. 1, 2, 2; fr. 20+, 15+, 4.3. – (1725).

CIL XV 1459 compl. et corr. ?   ☾   EX PR STATILI MAXIMI OP DO
                                                        VERATI MODES
                                                    nux pinea

Ex pr(aedis) Statili Maximi op(us) do(liare) Verati Modes(ti).

This example, which completes the version in *CIL* in the description of the form and *signum* and differs from it in the abbreviation DO instead of DOL, probably constitutes a correction rather than a variant of *CIL* XV 1459, which Dressel knew only from a single copy published by Fabretti and Marini. Since the stamp can be dated by type to the late Hadrianic period, the *dominus* should probably be identified with T. Statilius Maximus Severus Hadrianus, suffect consul in 115, who owned the *figlinae Brutianae* and *Macedonianae* at least as late as the year 134, rather than the consul of 144, T. Statilius Maximus, as Setälä has recently proposed (for the consul of 115, see Bloch, *BL*, 181; *AJA* 63 (1959), p. 236 n. 44, p. 237 n. 49; Steinby, *Cronologia*, 28–29; for the consul of 144: Setälä, *Domini*, 188–189; cf. 186–187). Setälä bases her identification on the absence of the cognomina Severus and Hadrianus in the latest stamps of Statilius Maximus, but the variance in nomenclature seems to reflect the practices of two different brickyards rather than a succession of *domini*: both versions of the name were used in the year 134, the shorter form at the *figlinae Macedonianae* (cf. *CIL* XV 288–289), the longer, with the cognomen Severus, at the *Brutianae* (cf. *CIL* XV 1453–1455; *S* 18; Taglietti, *RendAccLinc* 28 (1973), p. 313 no. 31 = Novum *CIL* XV *S* 18/41). Indeed, every stamp of Statilius Maximus that can be attributed with certainty to the *Brutianae* includes some form of one of the two cognomina. Furthermore, no stamps of Statilius Maximus can be dated to later than the first years of Antoninus Pius, nor is there mention after that time of either the *figlinae Brutianae* or the *Macedonianae,* both of which evidently ceased operation at the death of their last known owner (see Steinby, *Cronologia*, 60. The *figlinae Platanianae,* which belonged to Statilius Maximus in the year 125, are likewise thereafter unattested: cf. Steinby, *Cronologia*, 72). It is not absolutely certain to which of the brickyards owned by Statilius Maximus this stamp belongs, but the absence of both "Severus" and "Hadrianus" would seem to rule out the *Brutianae,* and Steinby's observation that it is closest in type to stamps of the *Macedonianae* is all but conclusive (cf. *Cronologia*, 29). The *officinator* Veratius Modes(tus) is not otherwise known.

105.   Sig. 11.6, 3.4; litt. 1.4, 1.4; fr. 25+, 13+, 3.7. – (1726).

CIL XV *S* 386   ▭   T·TETTIVS
                                              BARBARVS·F

  T. Tettius Barbarus f(ecit).

  The corners of the stamp are rounded. The same stamp has been found on a *mortarium* at Pompeii and can be dated by type to the last period of the city (cf. *CIL* XV *S* 554). T. Tettius Barbarus is not otherwise known, but according to Bloch he was "obviously related" to the [T.] Tettius Restitutus whose name was likewise stamped on a *mortarium* found at Pompeii (comment at *S* 555). A new stamp of L. Pollius Albinus, which Steinby has dated from a discovery *in situ* to the time of Caligula and which is similar in paleography, form, and formula to *CIL* XV *S* 386, suggests a more precise dating of the latter to approximately the same period (cf. *Largo Argentina*, p. 331 no. 147 (Tav. LXXIX, fig. 7)).

106.   Sig. 8.4+, 2.6–2.8; litt. 2.1–2.2; fr. 11+, 8+, 2.8–3.2. (1377).

CIL XV 1490b   ▭   M·VARGV

  M. Vargu(ntei).

  The mark of punctuation is round; the transverse stroke of A is pendent. Dressel saw only a rubbing of the stamp, from which the shapes of the A and the final V could not be positively determined. On this example the V, though fragmentary, clearly shows a serif, as does the A, though less distinctly: see the photograph and the comment in *CIL*. M. Vargun(teius) appears in brick stamps only in the two variants of *CIL* XV 1490, both of which can be dated on the basis of the paleography and form to the end of the Republic or the beginning of the Empire (for the homonym, cf. *CIL* VI 28328).

107.   Sig. 7.6, 2.7–2.8; litt. 1.2, 1.0–1.2; fr. 17+, 13+, 4.1. – (3069).

Novum *CIL* XV 1490/1   ▭   M·VARIENI
                                            ANTHVS·F

    M. Varieni Anthus f(ecit).

    The brick is broken and mended. The same Anthus appears as the slave of two Varieni in a stamp (*CIL* XV 1491) that Dressel knew from a fragmentary copy preserving just one of the *praenomina*, T(itus). The new stamp establishes beyond reasonable doubt what Bloch has already proposed, that the missing *praenomen* was M(arcus) (comment at *CIL* XV *S* 559). Titus is known only from *CIL* XV 1491, whereas slaves and freedmen of his brother (?) Marcus Iucundus and Marcus himself appear in several other stamps as well (*CIL* XV 964; *S* 559–562; cf. Hanslik, *RE* s.v. "Varienus" (1955), 382). In this respect, although *CIL* XV 1491 implies only joint ownership of a slave rather than an estate, the stamps of the Varieni suggest a relative chronology of their activity in the brick industry similar to that of the Domitii, Lucanus and Tullus: the business partnership of two brothers (or in this case, perhaps, of a father and son) was dissolved when one party left (possibly died), and the remaining member continued to manufacture as sole owner of the property (for the Domitii, see the comment at **80**; on *societas* in the brick industry, see Helen, *Organization,* 113–115; Steinby, *RE* Supplb. XV, 1519). The stamp recording the names of both Varieni, then, would predate those in which Marcus appears alone. All stamps of the Varieni can be approximately dated to the third quarter of the first century: *CIL* XV *S* 560–562 were stamped on *mortaria* found at Pompeii.

108.   Sig. 9.6, 4.1; litt. 1.5–1.6, 1.4–1.6; lin. 1, 1, 1; 19+, 12+, 3.6. – (1378).

CIL XV 1498 var. ?   ▭   T·VETTI ⚘
                                          FVSCI

    T. Vetti Fusci.

    This example differs from *CIL* XV 1498 in the decoration, which Dressel noted only after the second line and described as a poppy sprig with two flowers (*ramus papaveris flores duos habens*). At *Iscr. dol.* 1380a Marini printed a palm frond after the second line under the TI of VETTI, as did Gatti in *BullCom* 29 (1901), 134 and Dressel (following Fabretti) at *CIL* XV 2510, a *dolium* stamp which he evidently considered to be identical with the brick stamp (comment at 1498; cf. Bloch, *Indices*, 50). On the Kelsey example the palm frond seems to extend the height of both lines (see the photograph). The stamps of T. Vettius Fuscus should be seen in relation to several other brick and *dolium* stamps that suggest a family history of the Vettii involved in brick production during the second half of the first century: Fuscus, whose brick stamp is paleographically similar to a *dolium* stamp of P. Vettius Clemens (Coste, *RendPontAcc* 43 (1970–1971), p. 107 no. 15 (fig. 17)), may have been the son or freedman of the T. Vettius Clemens known from a *dolium* stamp found at Collatia (*CIL* XV *A* 209). Titus and Publius Clemens were perhaps brothers (both, apparently, were "C. f(ilii)") and the sons of C. Vettius Iucundus, whose brick stamp of the *figlinae Viccianae* can be dated by type to the middle of the first century (*CIL* XV *S* 617; cf. Steinby, *Cronologia*, 96 and n. 4). Neither the *gentilicium* nor the cognomen Clemens is uncommon, however, and the proposed genealogy is of course hypothetical.

109.   Šig. ± 9.4; orb. ?; litt. 1.2–1.3, 1.3; lin. 1, 2, 2; fr. 21+, 12+, 3.0. – (1379).

CIL XV 1504 compl.   ☺   [EX PR·] VIBI AIACIAN OP·DOL
                                      [T·] FL·CELER

     [Ex pr(aedis)] Vibi Aiacian(i) op(us) dol(iare) [T.] Fl(avi) Celer(is).

     CIL XV 1504 is completed in the description of the form. Dressel's theory that Vibius Aiacianus was eponymous owner of the *praedia Aiaciana* in the late Hadrianic or early Antonine period is more convincing than Helen's suggestion that the term "Aiaciana" was used in reference to the landed property of Vibius Aiacianus only after his death before it had been divided by his heirs; Steinby's new interpretation of *CIL* XV S 393, if correct, would establish beyond doubt that the brickyards were called the *Aiaciana* while Vibius was still alive (cf. *CIL* XV,1, p. 15; Helen, *Organization,* 71; Steinby, *Cronologia,* 25 with *LSO* 1078 (= S 393 compl.). Setälä's opinion is not entirely clear: see *Domini,* 201 and n. 2. For other *figlinae* named after persons, see Steinby, *RE* Supplb. XV, 1508). T. Fl(avius) Celer may be added to Steinby's and Setälä's lists (*loc. cit.*) of the *officinatores* of Vibius Aiacianus who are known only from a single stamp.

110.   Sig. 6.4; litt. 0.9–1.0; lin. 1, 2; fr. 11+, 9+, 3.2. – (1708).

Novum *CIL* XV 1534/5   [O]   C·VMBRI·S[   ]PHORI·
                                             palma

     C. Umbri *vel* Umbri(ci) S[ym]phori.

     The completion S[ym]phori coincides with the disposition of the text, which suggests that the stamp was round. Members of the Umbrius and Umbricius families appear frequently in stamps on various types of clay vessels (cf., for example, *CIL* XV 5768–5786; *CIL* X 8056, 386–391) but almost never in brick stamps; the single exception is Umbria C. f. Albina (= *PIR¹* V 598), whose name has been found stamped on a roof tile at Aufidena (*CIL* IX 6078, 176) and on a water pipe in Rome (cf. *NSc* 1895, 206). Although the presence of the *praenomen* Gaius in the nomenclature of both Albina and S[ym]phorus suggests a possible connection between the two, they cannot have been closely associated, since Albina is thought to have lived at the turn of the second and third centuries, whereas the stamp of S[ym]phorus can be dated on the basis of paleography and form to nearly a hundred years before that time.

111.   Sig. ± 7.8; litt. ?, (B 2.0, O 1.4, C 1.3), 1.3–1.4 (S 2.0); lin. – ; fr. 17+, 14+, 3.3–4.0. – (1382).

CIL XV 1550   O   [R S] P   lineis
                                  [O]F BOC   rectis
                                  S VII

     [R(ei) s(ummae)] p(rivatae) [o]f(ficina) Boc(oniana) s(tatio) (septima).

     For an explanation of the terminology used in stamps of the age of Diocletian, to which this example belongs, see Dressel and Mommsen in *CIL* XV,1, pp. 386–389 (see also Steinby, *RE* Supplb. XV, 1502–1503, 1528, with further bibliography). Although the meaning of the term *statio* is by no means clear, the fact that all but one of the stamps in which it occurs have been found in the Baths of Diocletian has led Bloch to conclude that the emperor introduced a new system of administration at the imperial brickyards expressly for their construction; the stamps that reflect this reorganization of the industry can therefore be dated to the years in which the Baths were built, 299–305/6, or shortly before (see *BL*, 311–312). For the identification of the *officina Boconiana* with the *figlinae Bucconianae* (= *Vocconianae*), which were first opened by the emperor Commodus, and for their location "in the region of ancient Ficulea and Fidenae", see Bloch, *Suppl.*, 114–116 (cf. Steinby, *Cronologia*, 29). The stamp has not been photographed previously.

112.   Sig. ± 6.8; litt. 1.1–1.2; lin. 1, 1; fr. 15+, 13+, 3.2–3.6. – (1383).

CIL XV 1560   O   [OF]F·BVC·PROCVLI   scriptura circulari

[Of]f(icina) Buc(oniana) Proculi.

Dressel noted the elongated mark of punctuation after OFF but made no mention of the ring in low relief at the center of the stamp (see the photograph, cf. **116** and **119**). According to Bloch, all stamps of the fourth century that are not found in the Baths of Diocletian must belong to the age of Constantine, since the brick industry of the capital fell into a rapid decline shortly after the center of government was transferred to the East in 330 (*BL*, 314, 316). Proculus was an *officinator* or perhaps merely a workman, but certainly not a *dominus,* for the *officina Buconiana* had belonged to the emperor's *summa res privata* since the time of Diocletian and had, in fact, never been in private hands (see the comment on the preceding stamp; cf. Bloch, *RM* 66 (1959), 200).

113.   Sig. 6.4; litt. 1.0; lin. 1, 1; fr. 12+, 10+, 3.9. – (1385).

CIL XV 1569a III   O   OFF S R F DOM ⚲   litt. cavis
                                                                ⚲

Off(icina) s(ummae) r(ei) f(isci) Dom(itiana).

= Steinby, *MemAccLinc* 17 (1974), p. 94 no. 1569a *timbro* II. Dressel noted that *CIL* XV 1569a had been impressed from various dies; Steinby recognized three varieties in the collections of the Antiquariums of the Forum and Palatine, two of which were subsequently discovered at Ostia and included in the edition of the Ostia stamps as *LSO* 1104 I and II. The publication of this example, which differs from the versions in *LSO* primarily in the orientation of the *signum*, completes the photographic record of the different stamps of *CIL* XV 1569a known from the Antiquarium collections. An example from the Lateran Baptistry that, unlike the other variants, is said to show two auxiliary lines at the center of the stamp may be another copy of *LSO* 1104 I, which has approximately the same diameter, or it may be fourth variant of *CIL* XV 1569a (cf. Steinby, *Battistero*, p. 117 no. 1). So many copies of 1569a have been discovered in the Baths of Diocletian—ninety-two by Bloch's count—that each of the three (or four) known varieties of the stamp may be presumed to have been represented by at least a few examples; all the stamps can therefore be dated with reasonable certainty to the years in which the Baths were built, 299–305/6, or shortly before (see Bloch, *BL*, 253, 310, and the comment at **111**). The *figlinae Domitianae* were in use from the Domitianic age until the reign of Septimius Severus, by which time they had passed into the hands of the emperor and had been divided into four sections (see Steinby, *Cronologia*, 37–40). That the *Domitianae* remained divided when they were reactivated in the time of Diocletian is shown by the fact that they belonged in part to the *summa res privata*, in part to the *summa res fisci* (cf. Steinby, *RE* Supplb. XV, 1528).

114.   Sig. 6.0; litt. 0.9–1.0; lin. 1, 2; fr. 14+, 11+, 2.8. – (1386).

CIL XV 1570b var.   O   TIMOD ꟻO Я Ƨ ꟻO

Of(ficina) s(ummae) r(ei) of(ficina) Domit(iana).

The new variant belongs to a series of fourth-century stamps from the *officina Domitiana* that differ from each other only in the abbreviation of the name of the brickyard: cf. *CIL* XV 1570b (DOMITI); *CIL* XV 1570a (DOMITIA); *LSO* 1107 = *CIL* XV 1571 var. (*) (= *S* 597 (*) ?) (DOMITIA I). Perhaps the different abbreviations represent different production units within the *Domitiana*, in which case the vertical stroke after DOMITIA in *CIL* XV 1571 and its variant, which until now has remained unexplained, might be interpreted as indicating the workshop where the stamp originated, or possibly the workman who made the brick (see the comment at *LSO* 1107). For the dating of this example to the age of Constantine, see the comment at **112**; for the *officina Domitiana*, see the comment on the preceding stamp (see also the comment at **111**).

115.   Sig. 6.8; litt. 1.5–1.6; lin. 1, 1; fr. 10+, 9+, 3.2. – (1387).

CIL XV 1572 (*)   O   ·MOᗡ·Я·Ƨ·ꟻꟻO

Off(icina) s(ummae) r(ei) Dom(itiana).

The diagonal stroke of R is nearly horizontal; the dot at the center of the stamp marks the spot where the handle was nailed to the die (see the photograph). Although this example shows no trace of the pentagram at the center of the stamp reported by Dressel, it has the same dimensions as a copy of *CIL* XV 1572 from the Lateran Baptistry and apparently came from the same die as the example photographed in Cozzo, Tav. XIX, fig. 56, where the pentagram is clear (cf. Steinby, *Battistero*, p. 125 no. 67). It is presumably not, then, a new variant but rather a copy of 1572 impressed from a dirty or defective die. For the dating of the stamp, probably to the age of Diocletian but possibly later, see Bloch, *BL*, 311 n. 230, 314. For the *officina Domitiana*, see the comment at **113**.

116.   **I.** Sig. 6.4; litt. 0.9–1.2; lin. 1, 1; fr. 31+, 26+, 2.2–2.8. – (1388).
             II. Sig. 6.1; litt. 0.8–1.1; lin. 1, 1; fr. 18+, 13+, 3.0. – (27122, 27121).

CIL XV 1574b   O   OF Ƨ P OF DOM

Of(ficina) s(ummae) p(rivatae) of(ficina) Dom(itiana).

Cf. *LSO* 1108 = *CIL* XV 1574a. The two varieties of *CIL* XV 1574b in the Michigan collection differ from each other primarily in dimensions and in the disposition of the text: see the photographs (cf. Dressel *ad loc.*: "Hoc sigillum signaculis diversis impressum extat"). All three examples show at the center a ring in low relief around a dot that marks the spot where the handle was nailed to the die (cf. **112** and **119**). For the dating of the variants of *CIL* XV 1574 to the age of Constantine, see the comment at **112**; for the *officina Domitiana*, see the comment at **113**.

117.   Sig. 8.8; litt. 1.2–1.4, 1.3–1.4; lin. 1, ?; fr. 20+, 14+, 4.2. – (1391, 2412).

CIL XV 1628   O   OFF·T· Q·AVGG·ET·CAESS·NN·   litt. cavis
                                           S R   linea recta   litt. extantibus

Off(icina) T(   ) Q(   ) (duorum) Aug(ustorum) et (duorum) Caes(arum) n(ostrorum) s(ummae) r(ei).

On the example photographed in Cozzo, Tav. XV, fig. 42, the outer line of text is well preserved but the letters in relief at the center of the stamp are missing, probably because of a defect in the die rather than a deliberate cancellation of the text (cf. Steinby's comment on *CIL* XV 1564 in *MemAccLinc* 17 (1974), 94). Since the term "figlinae" is attested only once in stamps of the fourth century and is otherwise replaced by "officina", the initials T(   ) Q(   ) in this example presumably stand for the name of the brickyard rather than that of the *officinator* (cf. Steinby, *RE* Supplb. XV, 1502; Bloch, *Indices*, 92). The stamp has been found in the Baths of Diocletian and can therefore be dated to the years of the tetrarchy 292–305 (cf. Dressel's comment at *CIL* XV 1564 and Bloch, *BL*, 311).

118.   Sig. 5.8; litt. 1.1; lin. –, 1; fr. 10+, 7+, 2.7. – (1392).

CIL XV 1634   [O]   OFF S R̲ [F TE]MP ♥
                                              ♥

Off(icina) s(ummae) r(ei) [f(isci) Te]mp(esina).

The stamp has been found in the Baths of Diocletian and can be dated to the years of their construction, 299–305/6, or shortly before (see Bloch, *BL*, 311–312). The *figlinae Tempesinae* are well attested for the Trajanic and Hadrianic periods, when they were still in private hands, but after the first years of Antoninus Pius there is no mention of them again until the time of Diocletian, when they reappear as part of the *summa res fisci* (for the early history of the *Tempesinae*, see the comment at **60**). The stamp has not been photographed previously.

119.   Sig. 5.4; litt. 1.1–1.4; lin. 1, –; fr. 23+, 13+, 3.2. – (1750).

CIL XV *S* 604   O   ꙅ Я M ꟻO

Of(ficina) M(   ) R(   ) S(   ).

Cf. Bloch *ad loc.*: "S is presumably the initial of a workman, as F in [*CIL* XV] 1654b stands for *For(tunati)* (1654a)." The letters M and R, then, should be the initials of another brickyard newly opened in the fourth century (unless they represent a carver's mistaken transcription of the abbreviation M<a>r(ciana): cf. *CIL* XV 1613–1614), although neither Bloch nor Steinby recognizes them as such (see the comment at **117**; cf. Bloch, *Indices*, 89–90; Steinby, *RE* Supplb. XV, 1502, 1510–1511. In his index to Marini's volume, Gatti lists the *officina M. R. For.* (cf. *CIL* XV 1654a): *Iscr. dol.*, p. 541). The ring in low relief at the center of the stamp is common in the age of Constantine, to which this example probably belongs (see the comment at **112**; cf. **116**).

120.   Sig. 5.2; litt. 0.8–1.0; lin. 1, 1; fr. 16+, 14+, 3.8. – (1394).

CIL XV 1655   O   O E S P T

O(fficina) E(   ) s(ummae) p(rivatae) T(   ) ?

Dressel did not understand this stamp of late date and printed the text in the order P T O E S;
Bloch listed it among the stamps "of so cryptic a character as to defy any attempt to force them into the
indices" (see the comment in *CIL* and Bloch, *Indices*, 6–7). Steinby has suggested the interpretation printed
above, in which E(   ) presumably stands for the name of a new *officina* and T(   ), perhaps, for that of a
workman (see the comment on the preceding stamp).

121.   Sig. ?; litt. 1.1; lin. 1, 1; fr. 15+, 11+, 2.2. – (1395).

CIL XV 1660a compl.   [O]   D Ͷ ΛRC[ADIO ΛV]Ꝺ

D(omino) n(ostro) Arc[adio Au]g(usto).

Dressel knew only a single copy of the stamp from which the shape of the N could not be positively
determined, apparently because of a defect in the die: see his comment and the drawing in *CIL*. This example shows clearly that the N was inverted (see the photograph). The stamp, which dates from the years
383–408, survives in only a few copies (3 or 4 ?), all discovered at Rome (see *SPASR* I, p. 57 no. 233 (Pl.
VIII, 8) and Dressel's remarks on *CIL* XV 1660b). Cf. Dressel *ad loc.*: "Mirum est in tegulis urbanis
mentionem fieri Arcadii in oriente imperantis."

122.   Sig. 5.3+, 2.4; litt. 1.6 (IL 1.8, I 1.1); fr. 13+, 6+, 2.8. – (1722).

CIL XV *S* 422 corr./var. ?   ▭   PAMPHILI·[FAB FIG·TEG]

Pamphili [Fab(i) Fig(uli) *vel* fig(uli) teg(ula)].

This fragmentary example probably constitutes a correction rather than a variant of the version in
the *Supplement*, where the I in ligature PHIL is not reported; but it may be a variant of the contemporary
stamp *LSO* 1003 (= *CIL* XV 1355 corr.) or even a new stamp of the freedman C. Naevius Pamphilus before
his manumission (cf. *LSO* 989 = *CIL* XV 1332 corr./var. (*)). Paleography and form point to a date in the
early first century (note the open P's). For the cognomen Figulus, see the comment at **100**.

123.   Sig. 9.3; orb. 3.6; litt. 1.2, 1.0–1.1 (TI, NI 1.2), 0.8–0.9; lin. 1, 2, 2; fr. 18+, 13+, 3.3. – (1707).

CIL XV *S* 459 (*) ?   ☻   [E]X FIG INT·ELLIANIS DE POR COR   sic
                                       PETIN ET APRONIAN   a. 123
                                              COS

[E]x fig(linis) Int{·}ellianis de por(tu) Cor(   ) Petin(o) et Apronian(o) cos.

= *CIL* XV 2393 (*) ? Since punctuation is found nowhere else in this example, the triangular mark
after INT is perhaps due to an irregularity in the surface of the brick rather than a mistake of the carver: see
the photograph. The *figlinae Intellianae* are known only from this stamp, which is "urban" in both form and

formula and which Steinby rightly locates between the stamps of the *figlinae Germ(anicianae)* (*CIL* XV 245–247; cf. *LSO* 250) and those of the *figlinae ab Isis* (*CIL* XV 248–255) (*Cronologia*, 45 and n. 9; cf. the comment at **85**). The unidentified *portus Cor(   )*, perhaps, as Bloch suggests, *Cor(nelii ?)*, is to be compared with the *portus Licini* and the *portus Parrae* and considered a storage place and center for the distribution of bricks (see Bloch's comment *ad loc.* and, for the *portus Licini*, the comment at **36**).

Novum *CIL* XV 2401/2 = **8.**

FRAGMENTS OF UNEDITED STAMPS

124.   Sig. 8.1+, 3.7–3.9; litt. 1.4, 1.3–1.4; fr. 12+, 16+, 3.0. – (1719).

Novum   ▭   protome Mer- ANT[
                             curii ? ds.   AEL[

Ant[imachus ?] Ael[i] ?

Steinby has tentatively identified the *signum* as a bust of Mercury, a conventional motif on stamps of the first century. A similar (perhaps identical) *signum—protome viri ds.*—appears in the same position on a roughly contemporary stamp datable to the early Flavian era (*CIL* XV 941; cf. *CIL* XV 1149 of the same man, after his manumission, where a bust of Mercury is figured at the right of the text). To judge from the character of the lettering, the use of the *signum*, and the form of the stamp, the new fragment might date from as early as the middle of the first century, but the proposed identification of the person named in the text suggests that it belongs to a slightly later period, closer to the Vespasianic age. For it seems that the first line recorded the name of a slave, Antimachus, who appears as a freedman, Aelius Antimachus, in a circular stamp approximately datable to the time of Domitian (*CIL* XV 778 = Cozzo, Tav. XX, fig. 60); the second line probably included an abbreviation of *servus* or of *fecit*, or perhaps the cognomen of the *patronus*, whose identity remains uncertain (cf. the comment at **127**). Several Aelii are known from stamps of the second half of the second century, but only one, Aelius Diphilus, is mentioned in a first-century stamp, where he appears as the owner of a slave Felix (*CIL* XV 1138; the same Diphilus recurs in *CIL* XV 1869, a circular fragment of uncertain date: cf. *CIL* XV *A* 77). Antimachus and Diphilus may well have been associated, but the *gentilicium* Aelius is too common to confirm any direct connection between the two (cf. also the Aelii Alexander and Silvanus, who worked as *officinatores* at the *figlinae Sulpicianae* of Domitia Longina during the late Trajanic and early Hadrianic periods: see Steinby, *Cronologia*, 90 and n. 1; Setälä, *Domini*, 109–110; cf. also Bloch's comment at *CIL* XV *S* 158).

125.   Sig. 8.0+, 2.6; litt. 1.6–1.7; fr. 12+, 10+, 3.1. – (2410).

Novum   ▭   ]ANTI·EPAPR

]anti Epapr(oditi) *vel* Epapr(ae) ?

Although the new fragment is similar in dimensions and paleography to a contemporary stamp of C. Calvisius Amarantus (*LSO* 736 = *CIL* XV 911d (*)), the missing text should probably not be restored Amar]anti: since the latter part of the stamp is best interpreted as an abbreviated cognomen, Epapr(oditus), or possibly Epapr(a), the preceding letters presumably form part of a *gentilicium*, perhaps [Sc]antius, or even Anti(stius) (for Scantius, cf. *LSO* 19 (= *CIL* XIV 5308, 34 (*)); for Antistius: *CIL* XV 2534; *A* 120; cf. also *A* 246. Both *gentilicia* occur with the cognomen Epaphroditus: cf. *CIL* VI 25978, 11899). It is perhaps possible to read an I in ligature A̱P̱I̱, in which case one might think of the *gentilicium* Papirius (cf. *CIL* XV 1356 = **100**; *LSO* 1088) or the cognomen Papirianus (cf. *CIL* XV 354), but the E immediately preceding would then be difficult to explain (see the photograph; the abbreviation *e(t)* does not occur in stamps). Paleography and form suggest a date in the first half of the first century.

126.   Sig. 4.2+, 2.2; litt. 1.5–1.6; fr. 14+, 7+, 3.1. – (1703).

Novum   ▭   ]MACE

       ] Mace( ).

Although the abbreviation *Mace( )* for *Mace(donianum* sc. *opus)* is found in several second-century stamps (*CIL* XV 289, 295–296), it does not occur in stamps of the first half of the first century, to which this example belongs, nor, apparently, were single-line rectangular stamps ever used at the *figlinae Macedonianae*. The brickyards were operating at least as early as the time of Claudius, but here the abbreviation probably stands for the cognomen, perhaps Macer or Macedo, of the only person named in the stamp (for the *figlinae*, see Steinby, *Cronologia*, 59–60; for the cognomina, cf. Kajanto, *LC*, 244; Solin, *Beiträge*, 145).

127.   Sig. 6.6+, 3.5; litt. 1.2, 1.2; fr. 21+, 14+, 6.2. – (2413).

Novum   ▭   SER·RVB.[
                                                       EVEMER[

       Ser(vili *vel* -gi ?) Rub[ ] Euemer[us] ?

After B in the first line is a vertical stroke, possibly part of an E or an R. According to the interpretation given above, the name of the slave, Euemer[us], would probably have been followed by an abbreviation of *fecit*: see Bloch's comment at *CIL* XV S 559 (*fec.*); cf., for example, *CIL* XV 1276 (*fecit*), 1491 (*fe.*), 1505 (*f.*) (for the spelling Euemerus, cf. *CIL* XI 6689, 95). But the reading of the second line is uncertain: the V may be an X; the E, less probably, an F; the R an M or an N (see the photograph). The text, then, might be interpreted: *ex f(iglinis) Me.*[ ], and one would think of the *dominus* L. Memmius Rufus, whose stamps have been dated to the turn of the first and second centuries (cf. *CIL* XV 1821; for Memmius Rufus, see Setälä, *Domini*, 149–150 and, on the probability of an identification with a known landowner, 10–11). The unidentified Ser( ) Rub( ) is perhaps the same as the Ser( ) Ru( ) known from a fragmentary stamp tentatively dated to the time of Nero (cf. Coste, *RendPontAcc* 43 (1970–1971), p. 107 no. 13 (fig. 15)): the paleography and form of this example suggest a date in the second half, perhaps the third quarter, of the first century.

128.   Sig. 7.1+, 2.3; litt. 1.3–1.7; fr. 12+, 11+, 3.1. – (2409).

Novum   ▭   ]ESCVИ

Given the uncertainty of the readings, any interpretation of the text must necessarily remain tentative, but since the elongated and slightly angular shape of the S, the semicircular form of the C, and the rounded ends of the letters all point to an early date, at the end of the Republic, the stamp is perhaps more likely to have recorded a *praenomen* and *gentilicium*—[. V]escun(iae), for example—than it is the *tria nomina*: [.]esc( ) Un( ), possibly [. P]esc(enni) or [. M]esc(ini) Un(ionis ?) (see the photograph; for "Vescunia", which occurs both as a *gentilicium* and as a cognomen, see Schulze, *LE*, 253, 319; for "Unio", see Kajanto, *LC*, 365).

STAMPS WITHOUT TEXT

129.   Sig. 5.5; fr. 15+, 11+, 3.1. – (1713).

    = *LSO* 1279 = *SPASR* I, p. 73 no. 341 (Pl. III, 5) = Van Essen, *Santa Prisca*, p. 263 no. 53–I/73 (Pl. LXXI, 1). The stamp consists of a ring enclosing a disc in relief on which four impressed heart-shaped figures, described as leaves in *SPASR* I and Van Essen, *Santa Prisca*, are arranged in the shape of a cross around a central depression. On this example the surrounding disc is incomplete, apparently because the die was applied unevenly.

130.   Sig. ± 3.0; orb. 1.3; fr. 6+, 5+, 3.2. – (1718).

    Two contiguous impressed stamps are preserved. One is fragmentary and apparently identical with no. **129** (cf. **131**); the other resembles an oblong orbicular stamp with the horns rounded off and is somewhat similar to the stamp published in *SPASR* I, p. 76 no. 381 (Pl. V, 15).

131.   Sig. ?; orb. 1.1; fr. 6+, 6+, 3.2. – (1716).

    The two stamps are similar to those described in the preceding entry, but they are not contiguous and the orbicular stamp is smaller and rounder.

132.   Sig. 4.3; fr. 10+, 7+, 2.7. – (1715).

    Two stamps are preserved. One, the diameter of which is reported above, consists of a narrow ring in relief surrounding a circular field on which an impressed (heart-shaped ?) figure adjoins a similar figure in relief. The other is impressed in the shape of an aryballos ca. 3.5 cm. long enclosing curvilinear figures in relief (cf. *SPASR* I, p. 75 nos. 368–369 (Pl. V, 1–2) for similarly shaped stamps without the figures in relief).

133.   Sig. 2.4; fr. 13+, 7+, 2.8. – (29535).

    The brick was given to the Kelsey Museum by E. B. Van Deman. Of the four stamps preserved, the incomplete example seems to be identical with the larger circular stamp, which consists of a cross impressed on an irregular disc in relief. The smaller round stamp is an incomplete circle ca. 0.9 cm. in diameter (cf. *SPASR* I, p. 81 no. 417 (Pl. VII, 15); Rodríguez Almeida, *Cuadernos de Trabajos de la Escuela Española de Historia y Arqueologia en Roma*, 12 (1969), p. 49 no. 57 (lám. I, fig. 2)). The fourth stamp is apparently formed by a squat T-shaped figure surmounted by a small curvilinear figure in relief.

134.   Sig. 4.0; fr. 15+, 8+, 2.7. – (93525).

Twelve small curvilinear figures in relief surrounded by two concentric rings are themselves arranged around a depression at the center of the stamp; nearby is a small circle 2.1 cm. in diameter. The fragment, which was given to the Kelsey Museum by a private collector, was evidently discovered at the end of the nineteenth century near Tivoli, two or three kilometers west of the site of Hadrian's Villa: a tag attached to the brick reads: "Colle Cesarano, Villa 21/4/99".

135.   Sig. ± 5.2; fr. 9+, 8+, 2.7. – (1717).

Five concentric rings at uneven intervals enclose a round depression at the center of the stamp; the third ring is narrower than the others; the outermost ring is 0.3–0.6 cm. inside the edge of the stamp (cf. *SPASR* I, p. 79 no. 402 and the comment *ad loc.* (Pl. VI, 6, 16, 18)).

136.   Sig. 3.0; fr. 10+, 9+, 3.0. – (1714).

= *SPASR* I, p. 79 no. 399 (Pl. VI, 15); see the comment *ad loc.* Three concentric rings at even intervals enclose a depression at the center of the stamp.

137.   Sig. 2.5; fr. 13+, 7+, 3.2. – (93523).

The brick was given to the Kelsey Museum by a private collector (cf. **134**). Six hollowed dots are arranged in a circle around a ring enclosing a disc in relief; seven larger hollowed dots form a T shape extending off the circular stamp.

MODERN OR FOREIGN STAMPS

138.   Sig. 11.4; litt. 1.3, 1.6; 2.0, 1.3, 1.5, 1.5, 1.5; 1.8 (O 1.2); 1.5, 1.9, 1.7, 1.7, 2.0; 1.6; fr. 21+, 18+, 5.1. – (2450).

$$\overline{\Theta \mathrm{Y}}$$

Ο   ΛΟΧΝΓ   lineis rectis
| ΝΑΟ.
\|ΝΡΙΒ
/

I do not understand this stamp, which Kelsey purchased at Athens. The first line evidently represents the abbreviation θ(εο)ῦ. The third line perhaps includes some form of the word ναός, or, if what seems to be traces of a letter at the end of the line is in fact merely encrustation on the surface of the brick, then the immediately preceding letter could easily be read as a theta, the parallel (?) strokes at the beginning of the next line could pass for an eta, and the text would be interpreted: Ἀθην(αῖς). Clearly, where so much of the text is in doubt, any hypothesis concerning the purpose of the stamp must necessarily remain tentative; but the interpretation of the first line, which is fairly certain, warrants at least the suggestion that the stamp belongs to the class of Greek brick stamps that marked the tiles destined for use in the construction of public temples and sanctuaries (see the useful list compiled by P. Mingazzini in *RendAccLinc* 25 (1970), 405–418, especially pp. 409–410, where the brick stamps of Athens are discussed).

139.   Sig. ?; litt. 1.3, 1.4; fr. 11+, 5+, 2.5. – (399).

formae   ]. ILIVSV[litt. cavis
irregularis   ]ＷＲＷ[

This example, which Kelsey purchased at Capri, is unlike any Roman brick stamp in paleography and form: the impressed border is irregularly contoured, the lettering is clumsily executed, and the die has left a flat, shallow impression in the surface of the brick (see the photograph). Steinby has suggested that the stamp is modern.

140.   Sig. ?; litt. 1.5–1.7 (Z, V 1.0); fr. 15+, 10+, 2.1. – (2851).

formae   ] VZVCR[litt. cavis
irregularis

Beneath the letters is an impressed curvilinear design. The stamp is paleographically similar to the preceding example and may have come from the same source, which Steinby has suggested is modern (see the photographs; note in particular the shape of the R's).

141.   Sig. 4.0; litt. C 2.7, P 3.2; fr. 11+, 9+, 2.2. – (2379).

C   P

litt. cavis

The letters are unusually large and have evidently been impressed from a modern die (see the photograph). A similar stamp is published in *SPASR* I, p. 56 no. 228 (Pl. II, 2); cf. the comment *ad loc.*

APPENDIX

Graphio Scripta

1.   Sesq. 42, 42, ?; litt. 12.7, 11.4, 11.4, 9.6, ?, 17.0. – (1059).

CCXXXX⎭

This brick from the De Criscio collection is broken into three pieces and held together by a wooden frame (see the Introduction, p. 6). The numerals were scraped into the wet clay in smooth shallow grooves, perhaps by a finger; the final stroke may not be part of the number, which probably indicates the total number of bricks made by a Campanian workman in a single day (240 ?). Similar graffiti on bricks found at Siscia in Pannonia suggest that workmen there were expected to produce at least 220 tiles a day during the brickmaking season, May to October (see Bloch, *BL*, 329–330; on the duration of the brickmaking season, see also Bloch, *CP* 39 (1944), 254–255). This example, inscribed on a *sesquipedalis*, supports Bloch's theory that the totals refer only to the production of the larger bricks, whereas a workman's daily output of *bessales* or *semilateres* would have been considerably greater (cf. *BL*, 330; on the interpretation of numbers inscribed on bricks, see also R. Tomlin in *BAR IntSer* 68 (1979), 233–236).

2.   Fr. 12+, 28+, 3.1; litt. M 1.8, I 2.1, P 2.3. – (2377).

M I P

M. I( ) P( ) ?

The brick, which Kelsey purchased at Argos, was marked before firing with three letters and a decoration, all inscribed with a stylus.

3.   Fr. 21+, 14+, 3.6. – (27128).

Before firing, the brick was inscribed with a stylus in deep cursive strokes ca. 3.5 cm. in length. The figure in the lower right corner of the photograph resembles a Greek theta—an oval enclosing a cross—but it is by no means certain that the marks were meant to represent letters (see the photograph).

71

CONCORDANCES

I. Variations with respect to *CIL* and *LSO*

I a. Unedited stamps (*nova*)

CIL X 8042	Cat. no.
60/1	**4**
64/5	**6**
76/7	**8**
94/5	**9**
98/9	**10**

CIL XV	
S 92/3	**31**
362/3	**32**
776/7	**69**
1204/5	**89**
1293/4	**94**
1336/7	**99**
1490/1	**107**
1534/5	**110**
unedited fragments	**124–128**

I b. Completions, corrections, and variants

N.B. Completions of punctuation, marked in the Catalogue with an asterisk (*), are not listed (see Concordance II a).

CIL VIII	*LSO*	Cat. no.
10474, 1 var.		**1**

CIL X 8042		
23 compl.		**3**

CIL XV		
20 var. 2		**13**
29b var. 1 II vel 2 II ?		**15**
S 25 corr.	= 112 compl.	**20**

CIL XV	*LSO*	Cat. no.
225 compl.		**27**
370	349 corr.	**34**
408d	387 compl.	**37**
507b var. ?		**45**
563c compl.		**51**
598a corr.		**58**
712	611 compl.	**65**
814 compl.		**71**
842b	701 corr.	**73**
858	709 corr.	**74**
1146b compl.		**85**
S 316 compl.		**86**
1284 compl.		**93**
S 342 corr. = *A* 59		**97**
1333 var. 2		**98**
1459 compl. et corr. ?		**104**
1498 var. ?		**108**
1504 compl.		**109**
1570b var.		**114**
1660a compl.		**121**
S 422 corr./var.		**122**

I c. Stamps identified with the same number in *CIL*

CIL XV	Cat. no.
29b var. 1 II vel 2 II ?	**15**
525c III vel IV	**47**
780b II vel III	**70**
1121c II	**83**
1203 II	**88**
1356 II ?	**100**
1569a III	**113**
1574b I–II	**116**

II

Concordance II comprises all stamps in the Michigan collection; the three graffiti presented in the Appendix are not included. In Concordance **II a**, a reference to *CIL* or *LSO* enclosed in parentheses indicates that the Michigan example could belong to any of the stamps listed. Completions of punctuation are marked with an asterisk only when the Michigan example shows additional punctuation not previously reported in *LSO* or elsewhere. In the last column, sources where illustrations can be found of stamps not photographed in this edition or in *LSO* are followed by an indication of the nature of the illustration: "ph." for "photograph" or "dr." for "drawing". The citations serve as a control on the identification of the Michigan examples and are intended merely to supplement the references in the two preceding columns. No attempt has been made to provide a complete list of the published photographs and drawings of stamps in the Michigan collection; further references can in many cases be found at the appropriate entries in the Catalogue or in *LSO*.

II a. *CIL* – Kelsey Museum accession no. – Catalogue no. – *LSO* – Other illustrations

CIL VIII	Kelsey acc. no.	Cat. no.	*LSO*	Other illustrations
10474,1 var. (*)	29541	**1**		
CIL X 8042				
19b	1064	**2**		
23 compl.	1128	**3**		
N. 60/1	1129	**4**		
61 (*)	1065	**5**		
N. 64/5	1066	**6**		
66	1127	**7**		
N. 76/7	1125	**8**		
N. 94/5	1124	**9**		
N. 98/9	1068	**10**		
99 (*)	1067	**11**		
CIL XV				
3	1138	**12**		
10	1139		27	
18	1140		31	
20 var. 2	1142	**13**	[36 = 20 var. 1]	
25a	1143	**14**		
26	93921		49	
29b var. 1 II vel 2 II ?	1141	**15**	[52 = 29b var. 2 I ?]	
(29b,c,d)	27001			
31	1144			
42 corr.	1145		= 76	
S 22	1745	**16**		
50a	27003		81	
50b	27004, 1147	**17**		
(50a,b)	1146			
50c	27005, 1148	**18**		

CIL XV	Kelsey acc. no.	Cat. no.	*LSO*	Other illustrations
(50a,b,c)	27002			
59a	1149		88	
59b	1150, 27006, 27007		89	
65	3065, 1151	**19**		
69	1152		101	
71	1153		104	
77 compl.	1735		= 110	
77 var.	1154		= 111	
S 25 corr.	1728	**20**	= 112	
79	1155, 2708		115	
81	1156	**21**		
84	1157		118	
87	1158	**22**		
88	1159		123	
S 33	2447			
97b	1160, 27009		134	
97c	2710		135	
97d corr.	1161		= 136	
98	27011, 1162		137	
104	1163		143	
115	29538		= 156 I	
118a	1164		158	
119a	1165		160	
(127, *A* 9)	1742			
134	1166, 27012	**23**	173	
S 43 (*)	1706	**24**		
157	27016			
162	1167, 27013, 27014, 27015		189	
163	1168		190	
144 corr.	1169, 27017		= 191	
169a	1170, 27018, 27019		197	
171	1171		201	
172	1172		202	
176	1173		205	
187	1174		213	
188	1175			
190b compl.	1176		= 214	
191	1177			
193	1178		216	
196	1179		217	
199	1180, 27020		220	
201	1181		222	
211	1182		228	
213	1183	**25**		
220 (*)	1184	**26**	233	

CIL XV	Kelsey acc. no.	Cat. no.	LSO	Other illustrations
221a	1185		234	
225 compl.	1186	27		
226	1187		239	
228	1188, 27021		241	
(230, S 184–185)	93924			
233	1190, 94941		244	
235	1191, 27022		245	
237b corr. = 237a	1192		247	
248	1193, 27023, 27024, 27025, 27026, 27027, 27028		253	
252	1194		256	
257	1195		266	
259	1196		268	
261a	1320		270	
271	1197		279	
273	1198		281	
286 (= S 69)	1199		293	
300	1200		299	
303a (*)	1201	28		
S 83 (*)	2407	29	306	
312	1202		307	
313	1203		308	
315	1204		310	
319	1705		313	
338	1205			Righini, Tav. XI, 1 (dr.)
354 (*)	1206	30		
(355 corr., S 92 corr.)	1207, 1208, 27029, 27030, 27033		(336, 337)	
S 92 corr.	1754, 1755, 1756, 27032		= 337	
N. S 92/3	27031, 1752	31	337/8	
S 93	1751		338	
359	1209		340	
361 corr.	1210		= 342	
N. 362/3	1211	32		
367 corr.	1212, 27034		= 346	
368 (*)	1213	33	347	
370	1214	34	349 corr.	
371a	1215, 27035, 27036		350	
375	1216		355	
(376b, 377c)	1217		(359)	
379	1218		361	
380 compl.	1219		= 362	
381	1220		363	

CIL XV	Kelsey acc. no.	Cat. no.	*LSO*	Other illustrations
393	1221		369	
398b	1222		377	
400	1223	**35**		
404	1224, 27037, 27038, 27039		383	
408a	1225, 27042, 27044		385	
408c	1226, 27045	**36**		
408d	1227	**37**	387	
(408a,c)	27041		(385)	
(408b,d)	27043		(386–7)	
(408b–e)	27040		(386–7)	
426	1228, 3066		396	
427a (*)	1229	**38**	397	
433 (*)	1230	**39**		
S 111	1730		405	
444	1231		406	
451	27046, 1232	**40**		
(454a,b)	1234		(416 I, 417 III)	
454b	1233		= 417 I	
S 115	27126	**41**		
458	1235			
468a	1236	**42**		
473	1237	**43**	433	
483	1238		443	
486a	1239	**44**	445	
(490–3; *S* 131–2)	29542		(447)	
495	1241, 27123		= 448 I	
	1240		(448 I–III)	
507b var.	1242	**45**		
515a	1243, 27047, 27048, 27049, 27050, 27051, 27052, 27053, 27054		457	
519	1245	**46**		
525c I	1246		= 462 I	
525c III vel IV	1720	**47**		
533a (*)	94940	**48**		
S 154	2411		474	
S 155	1723		476	
(549a,b,d)	27055			De Rossi, *Tellenae*, p. 39, fig. 54 (ph.) = 549a
(549b,c,d)	1247, 27056			
556	1248	**49**		
560	1249			
561	27057, 1250	**50**		
562	29524, 1251, 27058, 27059		486	

CIL XV	Kelsey acc. no.	Cat. no.	*LSO*	Other illustrations
563c compl.	29525	**51**		
563i	1253	**52**		
(563a,b)	27060, 27061			
(563a,b,i)	1252			
565e	1254	**53**		
(565a,e,k)	27062			
(566a,b,c)	1255			
574	27063, 1256			
575	1257	**54**		
583b I	27125			= *SPASR* I, p. 33 no. 111 I
583b II	27064			= *SPASR* I, p. 33 no. 111 II
583b III	1258			= *SPASR* I, p. 33 no. 111 III
585b	93923, 1259, 27065	**55**		
590	1260			cf. Steinby, *MemAccLinc* 17
591a	3062	**56**		(1974), 85
593 (*)	1261	**57**		
595a	1262, 27066			Righini, Tav. XVII, 1 (dr.)
(595b, 565)	1244			
(596b,c)	1263, 1264			Quilici, *Collatia*, p. 553,
598a corr.	1265	**58**		fig. 1186 (ph.) = 596b
599c	1266	**59**		
604a	93468			
604b	1267		497	
S 183	1397		514	
608a	1738		515	
609a	27067, 1268, 1739	**60**		
610	1269		512	
630b (*)	1270	**61**		
633a	1271	**62**	536 ?	
639	1272		544	
648b	1273	**63**		
650b	1274		553	
659b	1275		564	
659c	1276		= 565 I	
664a	27068, 1277	**64**		
680	1278			
683	1279		581	
692	1280		590	
695	1281, 27069		593	
696	81.1.3		594	
704a	1282		602	
704b	1283		603	
712	1284	**65**	611	
726	1285		623	
727	1286	**66**		
731b	1287, 27070,		627	

CIL XV	Kelsey acc. no.	Cat. no.	*LSO*	Other illustrations
	27071			
732	1288, 27072,		628	
	27073, 27074,			
	27075, 27076,			
	27077, 27078,			
	27079			
737	1289, 27080	**67**	633	
744	1290			Righini, Tav. XXII, 1 (dr.)
748	1291, 27081,	**68**		
	27082, 27083,			
	27084, 27085			
755	1292, 2559		641	
760a	1293		644	
762a	1294, 1295		646	
762b	29528		647	
763	1296		648	
773	1297, 27086		653	
N. 776/7	1724	**69**		
780b II vel III (*)	93922, 1298	**70**		
786	1299		659	
795a	1300		661	
811c	1301		682	
811d	1302, 27087		683	
814 compl. (*)	1396	**71**		
823	1303		= 693 I	
824	1304	**72**		
831	1305, 27088		698	
842b (*)	1306, 27089	**73**	= 701 corr.	
849	1307			Cozzo, Tav. XL VIII, fig. 158 (ph.) (cf. Taglietti, *RendAccLinc* 34 (1979), p. 204 no. 58)
858	1308	**74**	709 corr.	
862	1309		712	
887b	1310	**75**		
904a	1311	**76**	731	
910a	1312	**77**		
921a	1313			
931b	1314	**78**		
971 (*)	1315	**79**		
997	1744		771	
1000a	1316		773	
1000e ?	1317, 27091,		= 774	= De Rossi, *Tellenae*, p. 117, fig. 278 ?
	27092			
1002	1318	**80**		
1013	1319		780	
1027	1740		793	
1029c	1321, 27093		795	
1034 (*)	1322	**81**		

CIL XV	Kelsey acc. no.	Cat. no.	LSO	Other illustrations
1036	1323		800	
1055b	1324		814	
1059	1325		817	
1065	1326		823	
1068a	1327, 27094		826	
1075a	1328		834	
1077a = *S* 281	1329		836	
1081	1330		838	
1084a	1331		840	
1086	1332		842	
1089	1333		844	
1094c	1334		847	
1096b compl.	1335, 27095		= 855	
1096e	2448		857	
1097f	1336, 27096		860	
1102a	1337, 27097, 27098	**82**		
1102b	1338, 23469		866	
S 286 compl.	1702		= 867	
1106b	1339, 27099, 27100, 27101, 27102		874	
1108	1340		876	
1115a	1341, 1342		878	
1116c	1343		882	
1121c II	1344	**83**		
1137	1345		= 894 II	
1139	1736	**84**		
1146b compl. (*)	1130	**85**		
S 316 compl.	1398	**86**		
1181a	1346	**87**		
1189	29543, 1347		916	
1203 II	1348	**88**		
N. 1204/5	1349	**89**		
1220 (*)	1350	**90**	936	
1221a	1351		937	
1237	1126	**91**		
1248a	81.1.1	**92**		
1253c	81.1.2		955	
1253f	1352		957	
1257	1381			Cozzo, Tav. LXII, fig. 207 (ph.)
1269	1353		963	
1284 compl.	1354	**93**		
N. 1293/4	2408	**94**		
1301	1355		974	
1303	1356		976	
1314	1357, 27103		981	

CIL XV	Kelsey acc. no.	Cat. no.	*LSO*	Other illustrations
1317	1358	**95**		
1323a	1359	**96**		
S 342 corr. = *A* 59	1721	**97**		
1324 = *S* 343	1360		986	
1333 var. 2	1727	**98**		
N. 1336/7	27127	**99**		
1344	1361		996	
(1346b,c)	1362		(997)	
1346d	1363			Righini, Tav. XXXII, 2 (dr.)
1348a	1364		999	
1356 II ?	1365	**100**		
1368	1366, 27104		1011	
S 359c	1711		1020	
1387 = *A* 64	1709			*NSc* 1976, p. 315, fig. 54a (ph.)
1414	1367	**101**		
1418	1368		1039	
1423a	1369	**102**		
1430	1370		(1049 I-II)	
1432	1371, 27105, 27106, 27107		1051	
1434	1372, 27108, 27109, 27110, 27111, 27112, 27113, 27114, 27115		1052	
1442	1373	**103**		
1449a	1374		1059	
1449e	1375		1062	
1459 compl. et corr.?	1725	**104**		
1465a	1376, 27116		1069	
S 386	1726	**105**		
1490b	1377	**106**		
N. 1490/1	3069	**107**		
1498 var. ?	1378	**108**		
1504 compl.	1379	**109**		
1529a	1380		1097	
N. 1534/5	1708	**110**		
1538	1748		1101	
1550	1382	**111**		
1560	1383	**112**		
1566b	1384, 27117		1103	
1569a II	27118, 27119, 27120		= 1104 II	
1569a III	1385	**113**		
1570b var.	1386	**114**		
(1570c,d)	1757			
1572 (*)	1387	**115**		

CIL XV	Kelsey acc. no.	Cat. no.	LSO	Other illustrations
1574b I	1388	**116 I**		
1574b II	27122, 27121	**116 II**		
1587	1389			
1590	1390		1111	
1628	1391, 2412	**117**		
1634	1392	**118**		
1650	1393		1124	
S 604	1750	**119**		
1655	1394	**120**		
1660a compl.	1395	**121**		
1665b	94942		1126	
S 422 corr./var.	1722	**122**		
S 459 (*)	1707	**123**		

II b. *CIL* XV *S* – Kelsey Museum accession no. – Catalogue no. – *LSO*

CIL XV S	Kelsey acc. no.	Cat. no.	LSO
22	1745	**16**	
25 corr.	1728	**20**	= 112
33	2447		
43 (*)	1706	**24**	
69	1199		293
83 (*)	2407	**29**	= 306 (*)
92 corr.	1754, 1755, 1756, 27032		= 337
93	1751		338
111	1730		405
115	27126	**41**	
154	2411		474
155	1723		476
183	1379		514
281	1329		836
286 compl.	1702		= 867
316 compl.	1398	**86**	
342 corr. = A 59	1721	**97**	
343	1360		986
359c	1711		1020
363 corr. = A 64	1709		
386	1726	**105**	
422 corr./var.	1722	**122**	
459 (*)	1707	**123**	
604	1750	**119**	

II c. Unedited fragments

Kelsey acc. no.	Cat. no.
1703	**126**
1719	**124**

Kelsey acc. no.	Cat. no.
2409	**128**
2410	**125**
2413	**127**

II d. Stamps without text

Kelsey acc. no.	Cat. no.
1713	**129**
1714	**136**
1715	**132**
1716	**131**
1717	**135**
1718	**130**
29535	**133**
93523	**137**
93525	**134**

II e. Modern or foreign stamps

Kelsey acc. no.	Cat. no.
399	**139**
2379	**141**
2450	**138**
2851	**140**

II f. Unidentified fragments

In addition to four bricks that bear no mark whatsoever (Kelsey acc. nos. 1737, 2378, 27090, 27124), the Michigan collection includes twenty fragments of stamps that I have been unable to identify. Six of these, all fragments of circular stamps, consist of no more than five letters recording part of the phrase *opus doliare*:]OP·DOLI[(Kelsey acc. no. 1732);]OPVS D̦[(1747);]VS DOL E̦[(3064);]OP̦VS[(3067);]P DOLI[(3068); OP[(3070). Four others show only part of the formula *ex praedis*: EX PRAE[(Kelsey acc. no. 1729);]E̦DIS (1741);]EX P[(1746);]PRAE (3063). Two are fragments of circular stamps, one from the year 123, recording part of the consular dating formula:]ET PAET[(Kelsey acc. no. 1733; cf. *CIL* XV 28, 248, 791, *S* 113, etc.);]N̦ COS EX[(Kelsey acc. no. 1731; cf. *CIL* XV 552, 1042, 2394, etc.). Neither example is well enough preserved to confirm the apparent lack of punctuation between words. Another four fragments of circular stamps present common combinations of a few letters:]MITI[, presumably one of the stamps of the Domitii (Kelsey acc. no. 1704; cf. *CIL* XV 259, 261b, 269, 274, etc.);]BR[(Kelsey acc. no. 1710; cf. *CIL* XV 21, 27, 373, 374, etc.);]NA (Kelsey acc. no. 1743; cf. *CIL* XV 29c–d, 58, 662a, 1582, etc.);]̦ R (Kelsey acc. no. 1749; cf. *CIL* XV 61, 292, 540, 618, etc.). One badly worn example shows traces of the outermost line of a circular stamp:]P̦E̦ . . . . Q . . . (Kelsey acc. no. 1734). Another, completely illegible, has the form of a two-line rectangular stamp (Kelsey acc. no. 81.1.4). One from the age of Diocletian shows only the R[of *r(ei)* and the OF[of *of(ficinae)* (Kelsey acc. no. 1753; cm. 7.5). Finally, a circular stamp of which the text is illegible shows traces of the *signum* that Dressel describes as an eagle with wings spread, looking left (Kelsey acc. no. 1712; cf. *CIL* XV 157, 625, 628, 764, *S* 217, etc.).

INDICES

N.B. For an explanation of the format used in the Indices, see the Introduction, p. 12 and n. 36.

A. Nomina

The numbers after each listing refer to entries in the Catalogue.

A( ) V( ) ? 69
Ael[ius ?] 124
M. Aemilius Proclus **70**
  Aemilia Severa, c(larissima)
  f(emina) **38**
  [Ae]mili[a Severa] **39**
M. A(nnius) V(erus) **41**
Anti( ) Epapr( ) ? 125
Ant(onius) [L.] **72**
C. Aquilius Aprilis **32**
  Q. Aquilius Niger, *cos. a.*
  117, see ind. D
M. Arrius 2
  [A]rria Fadilla **21, 22**
  [Arria Fad(illa)] **20**
Q. Articuleius Paetus **73**
Asinia f. Marcell(i) **74**
L. Attius Cha[rito] **40**
Av( ) ? 69

L. Bellicius Sollers
  = L. Vellicius Sollers **75**
C. Brexius Senecio 3

Caecil(ia) Qu[inta] **54**
Caetennius, *see* Cet(ennius)
C. Cal[peta(nus) Favor] **76**
  Calp(etanus) Verna **26**
M. Calpurnius [P]hronimus **77**
  C. Calpurnius Spatalus **60**
Cal(vius) Crescens **66**
[C. Cass]ius C. f. Vet(us) **16**
Cet(ennius) M[ag]nio **88**
Cl(audius) Ionicus Caesaris
  Augusti l(ibertus) **71**

Clau(dius) Iu[ll(us)] **78**
[Cor(nelia) Ma]ll(iola) **44**
C. Cul( ) Dia(dumenus ?) **55**
Cuspius Melichrysus **79**

L. Domitius **83**
  Cn. Domitius Clemens **82**
  (Cn.) Domitius Tull(us) **80**
  Domitia Domitiani, *see ind. C*
  Domitia P. f. Lucill(a) **81**
  = Dom(itia) Luc(illa) **61**

[Fab(ius) Fig(ulus) *or* fig(ulus)] **122**
T. F(lavius) Amp(liatus) **46**
  (M.) Flavius Aper **85**
  [T.] Fl(avius) Celer **109**
  [F]l(avius) Corint(hus) **74**
  St. Fl(avius) M( ) ? 10
  Fla[via Do]mitil(la) **84**
Flavia Seia Isaurica, *see* Seia Isaurica

Gavia Amylla **31**
Q. Gra(nius) Apol( ) **86**

M. Her(ennius) Pol(lio) **87**

[C. Iulius Apoll]inaris **88**
  C. Iul(ius) Deyterius 6
  [C. Iulius Dio]scorus **89**
  C. Iulius Fortunatus **65**
  Iulius Fortunatus **102**
  [C.] Iulius L[upio] **20**
  L. Iulius Rufus **62, 64**

Iulia Augusta, *see ind. C*
Iulia Procula **63**
Iulia Saturn(ina) **90**

Q. Lepidius 7
= Lepidius **91**
L. Lurius Blandus **92**

(Statius) M(arcius) Ba[ssus] **21**
  [C. Mar]cius Pyrrichus **93**
[C. ?] Marsid[ius Am]aranth[us] ? 94
P. Mellutius Euvenus **48**
Mem( ) ? 127
Q. Muccius Asclep(iades) 8

L. Na[evius] **95**
  C. Naevius [Asc(lepiades)] **96**
  C. Naevius Bal(bus ?) **97**
  [C. Naevius P]hilomu(sus) 98
C. [N]unidius Felix **67**

O( ) V( ) *or* **Ov( ) ? 99**

P. P( ) B( ) **44**
Q. Papirius Figul[us] **100**
Q. Perusius Pude(ns) **33**
Sex. Plo(tius) Titia(nus) **68**
P(lotia) (Servilia) Is(aurica) **19**
  = Isaur(ica) **17, 18**
Sfx. [Po(mpeius)] **35**
T. Pom[ponius] **95**
Pop( ) Eun( ) **56**

[T. Rau(sius) Pa]mp(hilus) **19**
M. Rebulus Apronianus, *cos. a. 117, see ind. D*

M. R(utilius) L(upus) **13, 14**
  Rutilius Successus **23, 24**

Sabina Sabinilla **30**
C. Satrinius C[eler] **28**
Q. Sc( ) P( ) P( ) **101**
(Flavia) Seia Isaurica **102**
Ser( ) Rub( ) 127
(Ti.) Servilius Gelos **17, 18**
L. Servi(us) *or* Servi(lius) Fortunatus **103**
Statius, *see* (Statius) Marcius
  St(atius) Fl(avius) M( ) ? 10
  Stat(ia) Prim(ula) **61**
(T.) Statilius Maximus **104**

A. T(atius) L(aetus) 5, 11
T. Tettius Barbarus **105**
[T]re(bicia) (Tertulla) **47**

[(Q.) Valerius Cat]o **29**
M. Vargu(nteius) **106**
M. Varienus **107**
Vellicius, *see* Bellicius
Veratius Modes(tus) **104**
T. Vettius Fuscus **108**
Vibius Aiacian(us) **109**
(A.) Villius Alexander **50**
  C. Villius Cr[es(cens)] **57**
M. Vinic(ius) Fortunat(us) **101**
  [M. Vinic(ius) Pan]tag(atus) **52**
  = M. Vinic(ius) Pantg(atus) **51**
  = Vin(icius) Pa[n(tagatus)] **53**
(M.) Ulpius Anicetus **43**
  (M.) [Ulp(ius)] Ulpianus **45**
C. Umbrius *or* **Umbri(cius) S[ym]phorus 110**

[ ]antius Epapr( ) ? **125**
[ ]esc( ) Un( ) *or* [ ]escun( ) **128**

B. Cognomina

The numbers after each listing refer to entries in the Catalogue. A number included in square brackets indicates that the cognomen occurs also in Index A.

Agathobulus Domiti Tull(i) **80**
[Aga]thyrsus Aug(usti) lib(ertus) **42**
Aiacian(us) **[109]**
Alexander **[50]**

[Am]aranth[us] ? [94]
Amp(liatus) **[46]**
Amylla **[31]**
Anicetus **[43]**

Anthus M. Varieni 107
Antimachus Cl(audi) Ionici Caesaris
   Augusti l(iberti) **71**
   Ant[imachus] Ael[i] ? 124
Antoninus, *see ind. C; ind. D, a. 160*
Aper **[85]**
Apol( ) **[86]**
[Apoll]inaris **[88]**
Apol(lonius) Ant(oni) L. s(ervus) **72**
Aprilis **[32]**
Apronianus, *see ind. D, a. 117, 123*
Asclep(iades) [8]
   [Asc(lepiades)] **[96]**
Av( ) ? 69

B( ) **[44]**
Bal(bus ?) **[97]**
Barbarus **[105]**
Ba[ssus] **[21]**
Blandus **[92]**

[Cat]o **[29]**
C[eler] **[28, 109]**
Celsus, *see ind. D, a. 129*
Cha[rito] **[40]**
Clemens **[82]**
Corint(hus) **[74]**
[Cosmus] M. Her(enni) Pol(lionis)
   [ser(vus)] **87**
Crescens **[66]**
   Cr[es(cens)] **[57]**

Deyterius [6]
Dia(dumenus ?) **[55]**
[Dio]scorus **[89]**
[Do]mitil(la) **[84]**

Epapr(a) *or* **Epapr(oditus) [125]**
Euemer[us] ? 127
Eun( ) **[56]**
Euvenus **[48]**

Fadilla **[21, 22]**
   [Fad(illa)] **[20]**
[Favor] **[76]**
Faustus Domitiae P. f. Lucill(ae) **81**
Felix **[67]**
   Felix Fla[viaes Do]mitil(lae) **84**
Figul[us] **[100]**
   [Fig(ulus) *or* fig(ulus)] **[122]**

Fortunatus **[65, 102, 103]**
   Fortunat(us) **[101]**
Fuscus **[108]**

Gelos **[17, 18]**
Glabrio, *see ind. D, a. 152*

Hermes C. Cal[peta(ni) Favor(is)] **76**
Her(metianus) **34**
Homul(lus), *see ind. D, a. 152*

Ianuarius A. T(ati) L(aeti) 5
Ionicus **[71]**
Isaurica **[102]**
   Isaur(ica) **[17, 18]**
   Is(aurica) **[19]**
Iu[ll(us)] **[78]**

Jo( ) 53

L(aetus) [5, 11]
Larcio ? 85
Lucill(a) **[81]**
   Luc(illa) **[61]**
L[upio] **[20]**
Lupus **[15]**
   L(upus) **[13, 14]**

M( ) [10]
Mace( ) 126
M[ag]nio **[88]**
[Ma]ll(iola) **[44]**
Marcellus, *see ind. D, a. 129*
   Mar[cell(us)] **[74]**
Maximus **[104]**
Melichrysus **[79]**
Messal(la), *see ind. D, a. 115*
Modes(tus) **[104]**

Niger, *see ind. D, a. 117*

Ov( ) ? 99

P( ) P( ) ? **[101]**
Paetinus, *see ind. D, a. 123*
Paetus **[73]**
Pamphilus [Fab(i) Fig(uli)] **122**
   [Pa]mp(hilus) **[19]**
[Pan]tag(atus) **[52]**
   Pantg(atus) **[51]**

Pa[n(tagatus)] [53]
Pedo, *see ind. D, a. 115*
[P]hilomu(sus) [98]
[P]hronimus [77]
Pol(lio) [87]
Prim(ula) [61]
Priiscus Gaviae Amyllae 31
Proculus 112
    Proclus [70]
    Procula [63]
Pude(ns) [33]
Pyrrichus [93]

Qu[inta] [54]

[R]ab( ) 58
Rub( ) [127]
Rufus 59, [62, 64]

S( ) 119
Sabinilla [30]
Sagitta s(ervus) 72
Saturn(ina) [90]
Senecio [3]
Serapio *or* Serapion 9
Severus, *see ind. D, a. 141*
    Severa [38, 39]
Sollers [75]

Spatalus [60]
Stloga, *see ind. D, a. 141*
Successus A. T(ati) L(aeti) 11
    Success(us) [24]
    Sucessus [23]
    [Su]ces(sus) ser(vus) 27
S[ym]phorus [110]

T( ) 120
Titia(nus) [68]
Tull(us) [80]

V( ) ? [69, 99]
Verna [26]
Verus, *see ind. C; ind. D, a. 161*
    V(erus) [41]
Vet(us ?) [16]
Ulpianus [45]
Un( ) ? [128]
Urbicus 34

Yacinthus Iuliae Augusta(e) s(ervus) 4

Zosimus L. Iuli Rufi [62, 64]

[ ]aranth[us ?] [94]
[ ]escun( ) ? 128

C. Emperors and Their Families

The index of emperors is composed of two parts. In the first section the emperors and members of their families named on stamps are listed in chronological order. In the second section the anonymous references to emperors are followed by a letter or letters in parentheses indicating the emperor or emperors to whom the stamp may be assigned (see the summary of abbreviations below). Included in the first section are those stamps of the "anonymous" group for which the attribution is reasonably certain. The numbers after each listing refer to entries in the Catalogue.

Abbreviations

(C)	Commodus
(D)	Domitian
(DMCG)	Diocletian and Maximian and Constantius and Galerius
(H)	Hadrian
(M)	Marcus Aurelius
(P)	Antoninus Pius
(S)	Septimius Severus
(SA)	Septimius Severus and Antoninus (Caracalla)
(T)	Trajan

LIVIA AVGVSTI (*uxor*)
Iulia Augusta 4

DOMITIA DOMITIANI (*uxor*)
Dom(itia) Dom(itiani) **49**

HADRIANVS
Caes(ar) **32**

ANTONINVS PIVS
Caes(ar) n(oster) **33**

M. AVRELIVS
Aurel(ius) Caes(ar) n(oster) **27**
see ind. D, a. 161

FAVSTINA (*minor*)
F[aus(tina) A]us(usta) n(ostra) **35**
Faust(ina) Aug(usta) **23**
[Fa]us(tina) Aug(usta) **66**

L. VERVS
L. Verus Aug(ustus) **67**
see ind. D, a. 161

COMMODVS
Imp(erator) C[omo(dus)] Auc(ustus) **24**

SEPTIMIVS SEVERVS
D[ominus n(oster) Aug(ustus)] **25**

SEPTIMIVS SEVERVS et ANTONINVS
Aug(usti duo) n(ostri) **26**

ANTONINVS (Caracalla)
M. Aurelius An[to]ninus Aug(ustus)
n(oster) **36, 37**

DIOCLETIANVS et MAXIMIANVS
et CONSTANTIVS et GALERIVS
Aug(usti duo) et Caes(ares duo)
n(ostri) **117**

ARCADIVS
d(ominus) n(oster) Arc[adius Au]g(ustus) **121**

Caes(ar) n(oster) **43** (*T, H*), **33** (*P*)
C(aesar) n(oster) **68** (*H, P*)
Caes(ar) **32** (*H*)
Caesaris Augusti l(ibertus) **71** (*D-T*)

Aug(ustus) n(oster) **12** (*P, M*), **34** (*M, C*),
48 (*P, M, C*)
Aug(usti) lib(ertus) **42** (*T*)
d[ominus n(oster) Aug(ustus)] **25** (*S*)
Aug(usti duo) n(ostri) **26** (*SA*)
Aug(usti duo) et Caes(ares duo)
n(ostri) **117** (*DMCG*)

D. Consuls

Consuls whose names appear on stamps in the consular-dating formula are listed below in chronological order. The numbers after each listing refer to entries in the Catalogue.

a. 115
Mesal(la) et Pedone cos. 13

a. 117
Q. Aquilio Nigro M. Rebu[lo]
Aproniano cos. **14**

a. 123
Paetin(o) et Apronian(o) cos. **22**
Petin(o) et Apronian(o) cos. **123**
Paetin(o) et Apronia(no) **51, 52**
Paet(ino) et Apronia(no) cos. **81**
Paetino cos. **63**
Apr(oniano) et Paet(ino) cos. **44, 101**

a. 129
Marcell(o) et Cels(o) II cos. **45**

a. 141
Severo et Stloga [co]s. **90**

a. 152
[Gla]b(rione) et Homul(lo) cos. **42**

a. 161
Anso(nini) [III et V]ero II cos. 89

E. Figlinae, Praedia, Officinae, Portus

In the index of brickyards and estates, an *officina* of the fourth century that can be identified with a second-century brickyard of the same name immediately follows the latter in the list. The numbers after each listing refer to entries in the Catalogue.

[Bru]t[ia]na (*sc.* tegula) **15**
  Brut(iana) **13**
  Bru(tiana) **14**
[of]f(icina) Buc(oniana) **112**
  [o]f(ficina) Boc(oniana) **111**

[figlinae Ca]elian[ae] **16**
[Cae]pion(iana) (*sc.* tegula) **19**
  Cae(pioniana) **17, 18, 20**
f(iglinae) Can(inianae) **24**
  = fig(linae) Raninianae **23**

of(ficina) Domit(iana) 114
  off(icina) Dom(itiana) **113, 115**
  of(ficina) Dom(itiana) **116**

o(fficina) E( ) ? 120

figlin(ae) F[aurianae] **25**
  f[ig(linae) Fa]orian(ae) **26**
[fig(linae) Ful(vianae)] **27**

fig(linae) Intellianae de por(tu)
  Cor( ) **123**

of(ficina) M( ) R( ) 119
figlinae [Marcia(nae)] **28**

fig(linae) Negarianae **30**
ab Nept(uno) **31**

fic(linae) Ocea(nae) Mai(ores) **33**
  [fig(linae)] Oce(anae) Ma(iores) **34**

fi(glinae) Pontic[ula]nae **35**
de por(tu) Cor( ) **123**
port(us) Lic(ini) **36, 37, 61**
figl(inae) P[ubl]ilianae **38**

Q(uintanense) (*sc.* opus) **40, 41**

fig(linae) Raninianae, *see* f(iglinae)
  Can(inianae)

Sal(arese) (*sc.* opus) **44, 45, 47**
  [b(essalis) S(alaresis)] **46**
figl(inae) Sef( ) **48**
[figil(inae) St]rab(onianae ?) **29**
figli[nae] Sulp[icianae] **54**
  Sulpices(e) (*sc.* opus) **50**
  Sulp(icianum) (*sc.* opus) **51, 56, 58**
  Sul(picianum) **52, 53, 55, 57, 59**

off(icina) T( ) Q( ) **117**
Tempesina (*sc.* tegula) **60**
  off(icina) [Te]mp(esina) **118**
f(iglinae) Ter<e>nt(ianae) **61**
Tonneiana (*sc.* tegula) **62**

Vicciana (*sc.* tegula) **64**

F. Signa and Ornamenta

In the index of *signa* and *ornamenta*, the emblems and decorative elements found on stamps are listed in Latin in alphabetical order (the stamps without text, Catalogue nos. **129–137**, have not been included). The numbers in boldface type after each listing refer to entries in the Catalogue; a number enclosed in square brackets indicates that the decoration is not extant on the example or examples in the Michigan collection. After each Catalogue number the corresponding reference to *CIL* XV and the form of the stamp if other then the usual form ◒ are recorded in parentheses. Where not otherwise indicated, the *signum* appears at the center of the stamp.

anulus **112** (1560, ○), **116 I–II**
  (1574b, ○), **119** (*S* 604, ○),
  120 (1655, ○)

aquila alis expansis ss. respiciens
  duobus cornibus copiae insistit
  (*quasi in orbiculo*) **92** (1248a, ○)

73 (842b), **76** (904a), **82** (1102a), **93** (1284 compl., ◑)
ds. in principio versus **80** (1002)
ss. in principio versus **[18]** (50c), **[19]** (65, ◑),
  31 (N. *S* 92/3 ?, ◑), **43** (473), **62** (633a, ◑),
  64 (664a, ◑), **72** (842b), **76** (904a), **82** (1102a),
  93 (1284 compl., ◑)
ss. in fine versus **73** (858), **80** (1002)
stans (?) *in fine versus* **[100]** (1356 II ?, ▭)

Roma galeata ss. respiciens s. elata hastam, d.
  demissa sceptrum tenet; [pone eam sceptrum
  simillimum] **25** (213, ◯)

sagitta
  ds. in fine versus **62** (633a, ◑)
  ss. in principio versus **[62]** (633a, ◑)
  inversa in fine textus **30** (354)
  inversa inter verba **28** (303a, ▭)

sceptrum, *see* Roma

semina (?) quindecim (*supra textum*)
  86 (*S* 316 compl., ▭)

signum incertum (caduceus ?) (*in fine textus*)
  97 (*S* 342 corr., ▭)

sistrum (*ad d.*) **86** (*S* 316 compl., ▭)

situla (*ad s.*) **86** (*S* 316 compl., ▭)

sol ?, *see* discus

stellae duae sex radiorum, una in principio, altera
  in fine versus **64** (664a, ◑), **65** (712), **76** (904a),
  92 (*tertia inter verba*) (1248a, ◑)

testudo, *see* Mercurius

vas utrimque ansatum ds. iacens **48** (533a)

Victoria ss. respiciens, s. palmae ramum,
  d. elata coronam tenet; ad s. ara **37**
  (408d compl., ◯)

G. Notabilia Varia

The smaller indices are elaborated after the system established by Bloch in his *Indices* (pp. 93–103). The numbers after each listing refer to entries in the Catalogue.

1. *A distinction*
c(larissima) f(emina) **38**

2. *Abbreviations denoting the administration of* officinae *of the age of Diocletian (see the comment at* 111)
[r(es) s(umma)] p(rivata) **111**
S VII (= *statio septima*) **111**
of(ficina) s(ummae) p(rivatae) **116**
s(umma) p(rivata) 120
off(icina) s(ummae) r(ei) **115**
  of(ficina) s(ummae) r(ei) **114**
s(umma) r(es) **117**
off(icina) s(ummae) r(ei) f(isci) **113, 118**

3. *A legionary stamp*
leg(io) III̅ Au[g(usta) ?] 1

4. *A building in Rome*
castra praetori(a) Aug(usti) n(ostri) **12**

5. *Words pertaining to the production of* bricks
[b(essalis) S(alaresis)] **46**
bipidale dolia(re) 32
fa(cit) ? **88**
  <f>ec(it) ? **74**
  f(ecit) **73, 105, 107**
figlinae **16, 20** (feglinae), **28, 49, 54**
  figlin(ae) **25**
  figl(inae) **29** ([figil.]), **38, 48**
  fig(linae) **17, 18, 22, 23, 26, 27, 30, 33**, (fic.), **43, 63, 89, 123**
  fi(glinae) **35**
  f(iglinae) **19, 24, 44, 61, 75, 78**

6. Grammatica quaedam
a. *Single letters for double*

b. *Double letters for single*

c. *Variants of individual letters*

d. *A substituted ending*

e. *Irregular first declension*

f. *Syntax*

ex *with the accusative omitted in the second position:* **23** (*see the comment* ad loc.)

PLATES

1. *CIL* VIII 10474, 1 var. (*)   (29541).

2. *CIL* X 8042, 19b   (1064).

6. Novum *CIL* X 8042, 64/5   (1066).

4. Novum *CIL* X 8042, 60/1   (1129).

5. *CIL* X 8042, 61 (*)   (1065).

3. *CIL* X 8042, 23 compl.   (1128).

7. *CIL* X 8042, 66   (1127).

9. Novum *CIL* X 8042, 94/5   (1124).

8. Novum *CIL* X 8042, 76/7   (1125).

11. *CIL* X 8042, 99 (*)   (1067).

10. Novum *CIL* X 8042, 98/9   (1068).

12. *CIL* XV 3   (1138).

13. *CIL* XV 20 var. 2 (1142).

14. *CIL* XV 25a (1143).

15. *CIL* XV 29b var. 1 II vel 2 II ? (1141).

16. *CIL* XV *S* 22 (1745).

17. *CIL* XV 50b   (27004).

18. *CIL* XV 50c   (27005).

19. *CIL* XV 65   (3065).

20. *CIL* XV *S* 25 corr.   (1728).

21. *CIL* XV 81   (1156).

22. *CIL* XV 87   (1158).

23. *CIL* XV 134   (1166).

24. *CIL* XV *S* 43 (*)   (1706).

25. *CIL* XV 213   (1183).

29. *CIL* XV *S* 83 (*)   (2407).

26. *CIL* XV 220 (*)   (1184).

27. *CIL* XV 225 compl.   (1186).

28. *CIL* XV 303a (*)   (1201).

30. *CIL* XV 354 (*) (1206).

31. Novum *CIL* XV *S* 92/3 ? (27031).

32. Novum *CIL* XV 362/3 (*) (1211).

33. *CIL* XV 368 (*) (1213).

34. *CIL* XV 370 (1214).

35. *CIL* XV 400 (1223).

36. *CIL* XV 408c (1226).

37. *CIL* XV 408d (1227).

38. *CIL* XV 427a (*) (1229).

39. *CIL* XV 433 (*) (1230).

40. *CIL* XV 451 (27046).

41. *CIL* XV *S* 115   (27126).

42. *CIL* XV 468a   (1236).

43. *CIL* XV 473   (1237).

44. *CIL* XV 486a   (1239).

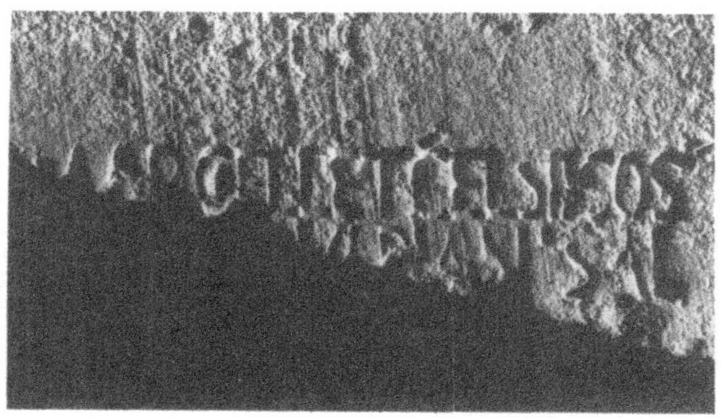

45. *CIL* XV 507b var. vel (*)   (1242).

46. *CIL* XV 519   (1245).

47. *CIL* XV 525c III vel IV   (1720).

48. *CIL* XV 533a   (94940).

49. *CIL* XV 556   (1248).

50. *CIL* XV 561   (27057).

51. *CIL* XV 563c compl.   (29525).

52. *CIL* XV 563i   (1253).

54. *CIL* XV 575   (1257).

53. *CIL* XV 565e   (1254).

55. *CIL* XV 585b (*)   (93923).

56. *CIL* XV 591a   (3062).

57. *CIL* XV 593 (*)   (1261).

58. *CIL* XV 598a corr. (1265).

59. *CIL* XV 599c (1266).

60. *CIL* XV 609a (27067).

61. *CIL* XV 630b (*) (1270).

62. *CIL* XV 633a (1271).

63. *CIL* XV 648b   (1273).

64. *CIL* XV 664a   (27068).

65. *CIL* XV 712   (1284).

66. *CIL* XV 727   (1286).

67. *CIL* XV 737 (*)   (27080).

68. *CIL* XV 748   (1291).

69. Novum *CIL* XV 776/7   (1724).

70. *CIL* XV 780b II vel III (*)   (93922).

71. *CIL* XV 814 compl. (*)   (1396).

72. *CIL* XV 824   (1304).

73. *CIL* XV 842b (*)   (1306).

73. *CIL* XV 842b (*)   (27089).

74. *CIL* XV 858 (*)   (1308).

75. *CIL* XV 887b   (1310).

76. *CIL* XV 904a   (1311).

78. *CIL* XV 931b   (1314).

79. *CIL* XV 971 (*)   (1315).

77. *CIL* XV 910a   (1312).

80. *CIL* XV 1002   (1318).

81. *CIL* XV 1034 (*)   (1322).

82. *CIL* XV 1102a   (1337).

83. *CIL* XV 1121c II   (1344).

84. *CIL* XV 1139   (1736).

85. *CIL* XV 1146b compl. (*)   (1130).

86. *CIL* XV *S* 316 compl.   (1398).

87. *CIL* XV 1181a   (1346).

90. *CIL* XV 1220 (*)   (1350).

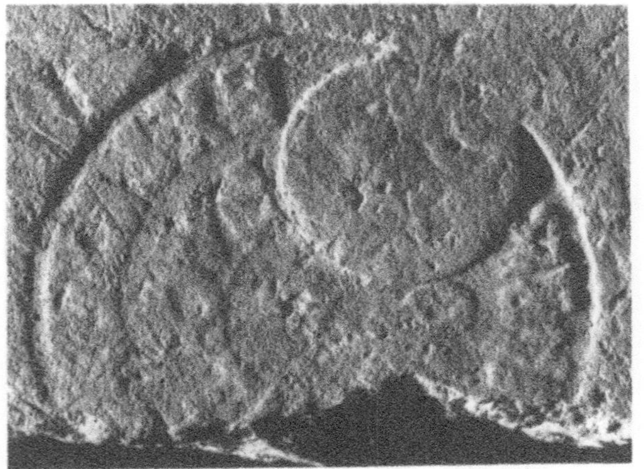

88. *CIL* XV 1203 II   (1348).

91. *CIL* XV 1237   (1126).

89. Novum *CIL* XV 1204/5   (1349).

92. *CIL* XV 1248a   (81.1.1).

93. *CIL* XV 1284 compl.   (1354).

94. Novum *CIL* XV 1293/4   (2408).

95. *CIL* XV 1317   (1358).

96. *CIL* XV 1323a   (1359).

97. *CIL* XV *S* 342 corr. = *A* 59   (1721).

99. Novum *CIL* XV 1336/7   (27127).

98. *CIL* XV 1333 var. 2   (1727).

100. *CIL* XV 1356 II ?   (1365).

101. *CIL* XV 1414   (1367).

102. *CIL* XV 1423a   (1369).

103. *CIL* XV 1442   (1373).

104. *CIL* XV 1459 compl. et corr. ? (1725).

105. *CIL* XV *S* 386 (1726).

106. *CIL* XV 1490b (1377).

107. Novum *CIL* XV 1490/1 (3069).

108. *CIL* XV 1498 var. ? (1378).

109. *CIL* XV 1504 compl.   (1379).

110. Novum *CIL* XV 1534/5   (1708).

111. *CIL* XV 1550   (1382).

112. *CIL* XV 1560   (1383).

113. *CIL* XV 1569a III   (1385).

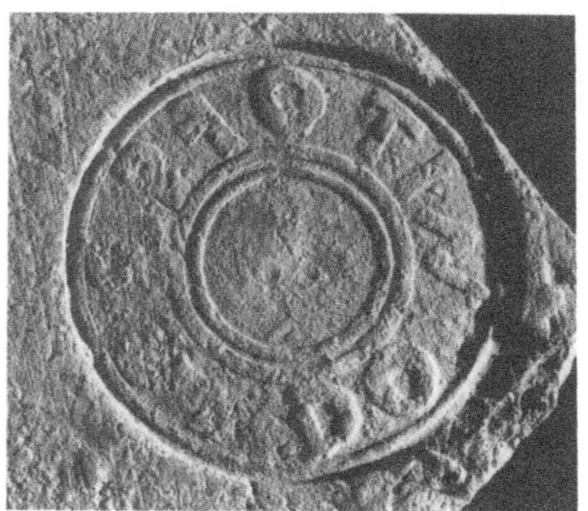

114. *CIL* XV 1570b var.   (1386).

115. *CIL* XV 1572 (*)   (1387).

116. I. *CIL* XV 1574b I   (1388).

116. II. *CIL* XV 1574b II (27122).

117. *CIL* XV 1628 (1391).

118. *CIL* XV 1634 (1382).

119. *CIL* XV *S* 604 (1750).

120. *CIL* XV 1655 (1394).

121. *CIL* XV 1660a compl. (1395).

122. *CIL* XV *S* 422 corr./var. ?   (1722).

123. *CIL* XV *S* 459 (*) ?   (1707).

124. Novum   (1719).

126. Novum   (1703).

125. Novum   (2410).

127. Novum   (2413).

128. Novum   (2409).

129. Sine textu   (1713).

130. Sine textu   (1718).

131. Sine textu   (1716).

132. Sine textu   (1715).

133. Sine textu   (29535).

134. Sine textu (93525).

135. Sine textu (1717).

136. Sine textu (1714).

138. Graecum (2450).

137. Sine textu (93523).

139. Huius aetatis ?   (399).

Appendix 1.   (1059).

140. Huius aetatis ?   (2851).

Appendix 2.   (2377).

141. Huius aetatis   (2379).

Appendix 3.   (27128).

Printed and bound by CPI Group (UK) Ltd, Croydon, CR0 4YY

09/07/2026

14916231-0001